Universe on a T-Shirt

Universe on a T-Shirt

The Quest for the Theory of Everything

Dan Falk

VIKING
CANADA

VIKING CANADA
Published by the Penguin Group
Penguin Books, a division of Pearson Canada,
10 Alcorn Avenue, Toronto, Ontario, Canada M4V 3B2
Penguin Books Ltd, 80 Strand, London WC2R 0RL, England
Penguin Putnam Inc., 375 Hudson Street, New York, New York 10014, U.S.A.
Penguin Books Australia Ltd, 250 Camberwell Road, Camberwell, Victoria 3124, Australia
Penguin Books India (P) Ltd, 11, Community Centre, Panchsheel Park,
New Delhi – 110 017, India
Penguin Books (NZ) Ltd, cnr Rosedale and Airborne Roads, Albany, Auckland 1310,
New Zealand
Penguin Books (South Africa) (Pty) Ltd, 24 Sturdee Avenue, Rosebank 2196, South Africa

Penguin Books Ltd, Registered Offices: 80 Strand, London WC2R 0RL, England

10 9 8 7 6 5 4 3 2 1

Quotation from *The Atom in the History of Human Thought* by Bernard Pullman (Oxford University Press 1998) is used courtesy of the publisher.

Printed and bound in Canada on acid free paper.

National Library of Canada Cataloguing in Publication

Falk, Dan, 1966-
    Universe on a T-shirt : the quest for the theory of everything / Dan Falk.

Includes bibliographical references and index.
ISBN 0-670-04335-4

    1. Physics—History—Popular works. I. Title.

QC173.7.F34 2002   530'.09   C2002-902908-2

Visit Penguin Books' website at **www.penguin.ca**

For my parents

Contents

Acknowledgements

I've always been interested in astronomy, physics, and the universe, but my passion for string theory blossomed after attending two conferences in Chicago in December 1996. One was held in memory of the Indian-born astrophysicist Subrahmanyan Chandrasekhar, and drew such luminaries as Stephen Hawking and Edward Witten. The other was the Texas Symposium on Relativistic Astrophysics, which, despite its name, is not always held in Texas. Since then I've become a bit of a science-conference junkie, attending symposiums and workshops across North America and becoming something of a "regular" at meetings of the American Astronomical Society and the American Physical Society. Stephen Maran, who handles media inquiries for the AAS, and Phil Schewe, who does the same for the APS, have given me numerous tips and contacts and answered hundreds of questions over the years.

My first major investigation of the themes in this book was carried out in 1997, when I put together a documentary called "From Empedocles to Einstein" for the CBC Radio program *Ideas*. That show was produced by Richard Handler, who helped focus many of my ideas. From that point onward, I had thought about writing a book—but the plan only crystallized after I got an e-mail from Jennifer MacTaggart of Penguin Putnam. Susan Folkins later helped steer me—and the book—in the right direction. I am also indebted to Rosemary Tanner for her tireless and invaluable help in editing the final manuscript, and to Joel Gladstone for his production skills.

Dozens of scientists and historians answered countless questions and, in some cases, agreed to look over portions of the manuscript; particularly helpful were the physicists Pekka Sinervo, Amanda Peet, Glenn Starkman, John Schwarz, and Michael Duff. Lengthy interviews with Edward Witten, Leon Lederman, Paul Davies, and John Barrow, which formed the heart of my *Ideas* documentary, also bring life to these chapters. On the humanities side, I benefited greatly from discussions with John Traill, Alexander Jones, James Robert Brown, and Dennis Richard Danielson. Owen Gingerich, an astronomer and historian of science, patiently granted interviews and responded to my numerous e-mail queries, while science writer Marcus Chown generously read several chapters. No doubt some errors remain; they are, of course, purely my responsibility.

Readers' comments are welcome and can be sent to universefeedback@hotmail.com.

Note to the Reader

Although I've tried to keep this book non-technical, there are occasions when distances, times, and other measurements come up. I have used metric units for distances, but those who are more familiar with Imperial units shouldn't be put off. A kilometer is a bit more than half a mile; a centimeter is a bit less than half an inch, and a millimeter is one-tenth of a centimeter.

In astronomy and cosmology, much larger distances arise and we must switch to "light-years." One light-year—the distance light travels in one year—is about 9500 billion kilometers or about 6000 billion miles.

Occasionally, we will also encounter very large or very small numbers. With large numbers, one can pile on strings of zeroes or refer to so many "billion billions"—but this soon becomes rather cumbersome; with small numbers ("billionths of billionths"), things become even uglier. The graceful solution, adopted by scientists everywhere, is to use *scientific notation*. Using this system, any number can be expressed as a "power of ten." With large numbers, an "exponent" stands for the number of zeroes. For example:

- one hundred = 100 = 10^2
- ten thousand = 10,000 = 10^4
- one billion = 1,000,000,000 = 10^9

and so on. With small numbers, the exponent becomes negative:

- one hundredth = 1/100 = 10^{-2}
- one ten-thousandth = 1/10,000 = 10^{-4}
- one billionth = 1/1,000,000,000 = 10^{-9}

Introduction

My ambition is to live to see all of physics reduced to a formula so elegant and simple that it will fit easily on the front of a T-shirt.

LEON LEDERMAN

The longed-for Theory of Everything promises to provide the final discovery after which all physics will become the refinement of its content, the simplification of its explanation....
Eventually, it will appear on T-shirts.

JOHN D. BARROW

The universe explained. Not a long-winded, highly technical explanation, but one that is concise, simple, and elegant. The "Theory of Everything" will explain the physical world we see around us—people and planets, cars and comets, sand and stars. It will explain the origin of everything in our universe, and describe its most basic components. And while it will likely be expressed through abstract mathematics, the ideas at the heart of the theory may turn out to be extremely simple—so simple, in fact, that the essence of the theory can be written on a T-shirt. This remarkable goal, suggested by the quotes on the preceding page from leading physicists on both sides of the Atlantic, sounds like a fantasy, something dreamed up after the Friday afternoon bull session moved from the faculty lounge to the local pub. Yet it is not a proposition made on a whim. It is a bold but very logical idea—a natural extension of what every physicist does every day. What they've been doing, in fact, since the dawn of science.

In the chapters to come, we'll explore this quest for simplicity in detail, tracing its evolution from ancient times, through the Scientific Revolution, to the provocative ideas of modern cosmology and particle physics. Let's start, however, not with history or with physics, but with a riddle. You're looking at a house across the street. You see a doctor enter, then a lawyer, and finally a priest. What is happening in that home? Before reading any further, take a moment to see if you can solve the riddle.

You probably figured out that the occupant of the house is terminally ill. First comes the doctor, who makes the diagnosis; then the lawyer, to settle the estate; and finally the priest, to administer last rites. What's so appealing about that solution, of course, is that it's so simple, so incredibly logical.

The situation described in the riddle is highly artificial, of course, but thousands of similar puzzles come along every day, confronting the scientist and non-scientist alike. Some are difficult; some are more straightforward. Suppose you walk into your living room and find one of the windows broken. Shards of glass lie on the carpet below the window. Nearby, a lamp is knocked over. Then, you see the crucial clue: a baseball, lying on the floor near the toppled lamp. The clues point overwhelmingly to one solution: someone must have hit a baseball through your window. Of course, you could come up with other explanations: maybe a burglar tried to get in through the window but gave up; meanwhile, the dog knocked over the lamp and dropped a baseball that he found in the yard. You don't have to be a scientist, though, to recognize that one solution is far more likely than the other.

When the answers are obvious, the problems seem trivial; when the answers are less obvious, the problems can take years or even centuries to solve. Even before modern science had developed, philosophers were thinking about the best ways to approach such problems. They realized that the simple, elegant solution was usually right. A medieval English monk named William of Ockham expressed this idea so succinctly that his name is now attached to the concept. The argument has become known as "Ockham's Razor."

Ockham's Razor is one of the most important "logical tools" that a scientist uses. Confronted with a bewildering array of data, the scientist looks for the simplest explanation for the observed facts. This is especially true in physics. The goal of the physicist has always been to *simplify*—to take a myriad of observations and explain them with as few laws and equations as possible. Physics is not philosophy, however, and the scientist has to take matters a few steps farther. Ockham's Razor is a useful guideline, but it's just the starting point. Every scientific theory must ultimately be

This Hubble Space Telescope photo reveals galaxies more than 10 billion light-years away. In the photo on the facing page, a magnetic field bends the paths of charged subatomic particles into distinct spirals. A successful Theory of Everything will apply to both realms, solving problems of the very large (in cosmology) and the very small (in particle physics).

tested by experiment. The theory has to predict specific results, and scientists have to perform laboratory tests to see if those results are obtained. A theory that doesn't agree with observational evidence—although a few die-hards may cling to it—will eventually be discarded. That approach, often labeled the *scientific method*, has yielded spectacular successes over the last 400 years, from the work of Galileo and Newton to the great revolutions of relativity and quantum theory in the twentieth century. It continues to enrich our view of the universe today.

In some branches of modern physics, however, experimental testing can be difficult—sometimes nearly impossible. Two fields are particularly challenging: high-energy particle physics, the search for nature's ultimate building blocks, and cosmology, the study of the origin and evolution of the universe itself. The two disciplines, which at first might seem unrelated, are in fact intimately linked. Particle physicists explore how matter and energy behave under extreme conditions, usually by smashing particles into one another in giant accelerators. Equally extreme conditions prevailed in the early universe, the realm studied by cosmologists.

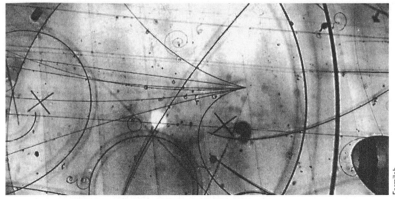

Fermilab

In the first split second after the big bang—the colossal explosion that gave birth to our universe some 15 billion years ago—the cosmos was so hot and dense that the various forces seen in nature today are thought to have acted as one. At that early time there was just a single force, a force which could be described by a single theory. Particle physicists and cosmologists are both struggling to understand that theory—the Theory of Everything.

Today, a great deal of excitement is focused on "string theory," which describes all the known particles and forces in terms of tiny loops of string. It's not as crazy as it sounds. With certain modifications, string theory may turn out to be our best hope for a unified theory of physics. Of course, we have no way to re-create the energy levels of the big bang in the laboratory; even today's largest particle accelerators cannot come close. In other words, we have no direct way of testing string theory. Where, then, does the theorist turn? One possibility is to test certain parts of the theory, to see if it's "on the right track," by checking for consistency with other, more strongly established theories. Another strategy is to heed the advice of Ockham and look for simplicity. These days, the science may seem esoteric and the equations complex—but the underlying ideas are often startlingly simple, even beautiful.

Driven by the search for concise, elegant explanations, physicists pursue theories that explain the widest range of physical phenomena. If possible, they test those theories in the laboratory; if not, they test them on the blackboard and on the computer. Often, they succeed. Each great revolution in the physical sciences, from Galileo and Newton to Maxwell and Einstein, has resulted in a new, simplified view of the universe; each has yielded an elegant description that can be expressed in a few simple equations.

In telling this story, I will also try to keep things simple. There will be no talk of "non-Abelian gauge theories," "renormalization," or "nonperturbative methods." Of course, we will still have to stay alert—we'll take an introductory tour of relativity and quantum mechanics, and we'll explore the world of strings, black holes, and hidden dimensions. But we'll never go any deeper than necessary to tie each discussion to our theme of simplicity and unification. And, though we'll occasionally talk *about* math, we won't actually *do* any math.

To truly understand this quest, however, we also need a sense of its history. We need to explore its origins, and its gradual evolution, beginning in the days when philosophy and science were one and the same. This book is a history of ideas, but it is also a story about people. Some were passionate experimenters, some were brilliant theoreticians, and a few were just plain lucky. But all were drawn by the quest for simplicity. All of them reached out for the holy grail of physics—the Theory of Everything.

Shadows and Light

The Greek World and the Beginning of Science

*Zeus, father of the Olympians, has made night
out of midday, hiding the bright sunlight,
and...fear has come upon mankind.*

ARCHILOCHUS, SEVENTH CENTURY B.C.

By mid-afternoon, the shepherds knew something was wrong. The sheep and goats were bleating and the birds were crowing as if it were evening—but it was still only late afternoon. Sunset was two hours away, yet the light was fading and the air was cooling. The light was also changing color, mimicking the orange of twilight before turning an eerie silvery-gray. And, though it sounds impossible, the shadows were becoming weaker and yet more crisp at the same time. The sun was disappearing.

The day was the 28th of May. By our calendar, the year was 585 B.C.; for the Greek settlers in the province of Ionia, in Asia Minor, it was the fourth year of the 48th Olympiad. In the towns along the coast of the Aegean Sea (today, part of Turkey), thousands of people stopped what they were doing to gaze at the spectacle unfolding in the sky. Though most of them did not know what was happening, some of the elders, and those who had traveled or had heard stories, understood that this was an *eklipsis*—an eclipse. It was, to be precise, a solar eclipse—the mid-day darkness that comes when the moon passes directly between the sun and the earth.

The eclipse also darkened the plains to the east, near the River Halys, where a great battle was raging. The Lydians, who lived just inland from the Greeks, were confronting the invading Medes, whose homeland was south of the Caspian Sea. We know about the battle from the writings of Europe's first historian, the Greek scholar Herodotus, who lived in the fifth century B.C. He tells us that the war had been waged for five years already, "during which both the Lydians and the Medes won a number of victories." He says that "after five years of indecisive warfare, a battle took place in which the armies had already engaged when day was suddenly turned into night"—a reference, historians agree, to the eclipse of 585 B.C. The two warring parties, viewing the eclipse as a sign of the displeasure of their gods—or perhaps simply seeing that an endless series of indecisive battles was doomed to bankrupt both empires—stopped fighting and quickly negotiated a peace treaty. The two leaders took an oath, Herodotus tells us, which was similar to that of the Greeks "but for additional confirmation they make a shallow cut in their arms and lick each other's blood."

The eclipse clearly caught the warring Lydians and Medes—and most of the Greeks—off guard. But a man named Thales (pronounced THAY-leez) was not so surprised. Herodotus tells us: "This change from daylight to darkness had been foretold to the Ionians by Thales of Miletus, who fixed the date for it in the year in which it did, in fact, take place."

The account given by Herodotus is, of course, open to scholarly debate. But if we take it at face value, then the eclipse of 585 B.C. marks a turning point in the story of human civilization. Rather than attributing the eclipse to the whims of the gods, Thales saw it as an event that happened for a logical reason, as the result of natural forces. Today's philosophers call such a view "naturalistic"—explaining things in terms of natural, rather than divine, laws. They might label Thales a "materialist"—someone who seeks explanations in terms of material forces and causes. (I'll try to avoid using the adjective "materialistic," as it makes us think of yuppies accumulating DVDs and driving SUVs.) Thales not only recognized eclipses as recurring natural phenomena, he also knew enough about their nature to predict when they would occur. (Compare this view to the reaction of the poet Archilochus, cited in the epigram at the start of this chapter. Archilochus placed the responsibility for that earlier eclipse—probably that of 648 B.C.—squarely at the feet of Zeus.)

It may be a slight exaggeration to label Thales' eclipse, as some historians have, as the birthday of Western science. But the approach of Thales and his fellow Greeks was certainly new. For the first time, people looked at the great variety and change displayed by nature and saw not chaos but an underlying order. And they committed themselves to using logic and reason to discover that order. In this new age of intellectual curiosity, people asked not just "when" and "where," but also "how" and "why."

The Birth of a Notion

The explosion of learning and curiosity that led to this new outlook had many roots. Greek ideas about politics and society—ideas that would eventually give rise to the first democracies—led to dissatisfaction with dogma and unquestioning obedience and a new respect for independent thought. A rise in literacy fostered the spread of ideas. Geography, too, played a role. Surrounded by the sea, Greek cities were influenced by the flow of goods and ideas from far-off lands, from the established civilizations of the Middle East to the nomadic tribes living to the north and west. By the time of Thales' eclipse, his hometown of Miletus was the cross-roads of the eastern Mediterranean, a focal point where eastern and western ideas mingled and new ideas were born. True, the average life span was just 35 years, but there was at least the hope that it would be 35 prosperous years. This prosperity, with the leisure time it fostered, was crucial for the intellectual growth that followed. ("Leisure," as the philosopher Thomas Hobbes said two millennia later, "is the mother of philosophy.")

The Greeks also knew good ideas when they saw them. Learning skills from earlier settlers in the region, they began farming the land, making tools and coins from metal, and using a written alphabet based on that of the Phoenicians. They grew grapes and olives, raised sheep for wool, and became experts in crafting pottery. But the Greeks didn't simply copy the locals. At their hands, the existing culture was transformed into something novel, reflecting the Greek spirit for adventure and innovation. As historian Thomas Goldstein writes, the Greeks "inherited from their nomadic forebears a natural vigor and independence of mind." Or, as Herodotus put it, "the Greek race was marked off from the barbarians as more intelligent and more emancipated from silly nonsense"—though we should bear in mind, of course, that Herodotus was a Greek himself.

Whatever ignited the Greek breakthrough does not seem to have taken hold in the other great civilizations of the region. The science of the Egyptians, for example, is usually seen as being purely technology-driven. The Egyptians were proficient architects and engineers—witness the pyramids—but they were more interested in specific practical applications than in the underlying science. The classic example of this is their ability to predict the annual flooding of the Nile based on the rising of the bright star Sirius. We might call it a scientific breakthrough, but, scholars argue, it was driven by the demands of agriculture and economics, not by a quest for astronomical knowledge.

The Babylonians, meanwhile, had a sophisticated calendar and reams of star charts; they, too, could predict eclipses. And there's no doubt about their skill in mathematics: they were masters of arithmetic, they understood the idea of the square root, and they had a good approximation for π (pi), the ratio of a circle's circumference to its diameter. They also knew of the Pythagorean Theorem at least a thousand years before Pythagoras gave his name to it. But the Babylonians, like the Egyptians, seemed to have been driven by immediate practical concerns.

That's not to say that these other cultures didn't contribute to the Greek achievement. Indeed, the Greeks acknowledged in their writings that they borrowed heavily from their neighbors. Much of their mathematics and astronomy came from Egypt and Babylon. The Greeks probably got the idea of dividing their days and nights into 12 parts from the Babylonians, and likely built their first sundials based on a Babylonian model. In fact, Thales was said to have exploited the Babylonian knowledge of eclipses in making his famous prediction—although historians now dispute the matter. (Worse still, the eclipse prediction may simply be a legend; in those days, great feats were often attributed to great men who just happened to be living at around the same time. But then, all such stories from that age are shrouded in mystery—and the tale is much too good to pass up.)

While we often describe these early Greeks as the first scientists, they didn't abandon their religious beliefs to make room for their new naturalistic ideas. The Greeks, in fact, had an abundance of gods. But their ideas about religion do seem to have shifted as this new way of thinking took root. The gods of their ancestors—the gods described by the great poets Homer and Hesiod—were intimately involved in human affairs. The lives of men and women were ruled by the arbitrary moods and desires of those gods, and events on the earth—whether good or bad—owed their origins to the whims of the deities. Every flood, every famine, every storm, every harvest: all were seen as their handiwork.

Over the years, however, the ideas of Thales and his followers—historians call them the Presocratics—began to spread. As this new philosophy blossomed, people had less and less need to attribute nature's spectacle to unknowable heavenly forces. These early thinkers "invented the very idea of science and philosophy," writes historian Jonathan Barnes:

> They saw the world as something ordered and intelligible, its history following an explicable course and its different parts arranged in some comprehensible system. The world was not a random collection of bits, its history was not an arbitrary series of events. Still less was it a series

of events determined by the will—or the caprice—of the gods. The Presocratics were not, as far as we can tell, atheists: they allowed the gods into their brave new world....But they removed some of the traditional functions from the gods. Thunder was explained scientifically, in naturalistic terms—it was no longer a noise made by a minatory Zeus ...the Presocratic gods do not interfere with the natural world.

This new Greek ideal was embodied in their word *logos*. Although our word "logic" comes from it, logos has no exact English translation; it encompasses both description and explanation. Sometimes read as "statement," "principle," or "law," logos was at the heart of the early Greek experiments in philosophy. When the Greeks searched for logos, they wanted not just any kind of answer, but the most far-reaching, authoritative answer they could achieve.

Thales: The Man from Miletus

Besides his prediction of the 585 B.C. eclipse, what do we know about Thales of Miletus, the man who supposedly brought Greece out from intellectual darkness? Thales (*c.*620–*c.*550 B.C.) may have written nothing of his own ideas, certainly nothing that survived, though later philosophers describe him as one of the Seven Sages of ancient Greece. Aristotle, writing two centuries later, called Thales the "first founder of this kind of philosophy"—the philosophy rooted in this new, materialist outlook.

According to Herodotus, Thales once diverted a river, allowing the king of Lydia to lead his army across. We're also told he was an able politician, helping establish the local government. We also know he traveled in Egypt, where he's said to have calculated the height of the pyramids based on the lengths of their shadows. (While in Egypt, he also may have witnessed the eclipse of 603 B.C., 18 years before the eclipse in his homeland.) Thales was

The Greek philosopher Thales was among the first to speculate on the make-up of the natural world.

also a mathematical prodigy, establishing four fundamental theories of geometry.

But we remember Thales today for his bold attempts to forge a philosophy of nature. Most famously, he said that *the universe is made of water.* For centuries, scholars have debated just what he meant by that. Did he mean that everything literally *is* composed of water? More likely, historians say, he believed that all things originated *from water* in the remote past. Thales may have been influenced by everyday observations of nature, noticing that living things require water for nourishment, and by the ever-present Mediterranean that surrounded the Greek world. Indeed, he believed that the earth itself rested on water. And he must have noticed that water, unlike other substances, routinely exists in three different forms—solid (ice), liquid, and gas (steam or water vapor), each of which has very different physical properties. Not everyone was swayed by Thales' reasoning. A later Milesian, Anaximenes (*c.*540–*c.*475 B.C.), proposed that everything was made of air. Heraclitus of Ephesus (*c.*540–*c.*450 B.C.) said that all was fire.

Each of these early thinkers was asking the same question: what is the world made of? The Greek philosopher-scientists, for the first time, were brash enough to suggest answers. Today, we might laugh at the child-like simplicity of their ideas, but it would be a mistake to dismiss them. It is not the *conclusions* these early thinkers arrived at that are important; it is the fact that *they dared to ask such questions.* No longer satisfied by myth, they turned to a philosophy of nature that would evolve into what we now call "science." And the next two centuries would bring even greater daring, as the Greeks searched for the ultimate physical explanation—the logos—for the natural world.

Empedocles Finds His Element

Empedocles (*c.*493–*c.*435 B.C.) was born in Acragas, a Greek colony on the island of Sicily. Here, in the shadow of Mt. Etna, Empedocles earned a reputation as a politician, an orator, and a poet; he may even have offered his services as a physician. Of his writings, two lengthy poems—or at least fragments of them— have survived. Among other things, they demonstrate his enormous ego. At one point, he even describes himself as divine:

> An immortal god, no longer mortal, I travel, honored by all, as is fitting, garlanded with bands and fresh ribbons. Whenever I enter a thriving town I am revered by men and women. They follow me in their thousands....

One legend says that he met his demise by throwing himself into a volcano, perhaps to prove that he really was immortal. Certainly he believed in reincarnation, viewing birth and death as illusions. Before his alleged plunge into Mt. Etna, however, Empedocles spent a great deal of time thinking about the natural world. He tried to explain why the earth is spherical and why the sea is salty. He may have even guessed—he could not have measured it—that light travels at a finite speed.

Most importantly for our discussion, it was Empedocles who first put forward the idea of the *elements*. He concluded that everything in the universe was made up of four kinds of material: *earth, air, fire,* and *water*. He believed that these basic elements— or "roots," as he called them—gave rise to everything seen in the natural world.

"Son, there are five basic elements; earth, air, fire, water, and mutual funds."

From these [elements] grow all things that ever were and are and will be: Trees, and men and women, and birds and beasts, and the fish nourished in salty water, as well as the long-lived gods, honoured above all. For they [the elements] are always themselves, but running through each other, they take on various forms and shapes....

Empedocles compared his vision to that of a painter, who can produce any color by mixing red, blue, and yellow in the right proportions. Nothing is created or destroyed, he argued; everything is simply the mixing and re-mixing of these basic, eternal elements. He also had a theory about what governed the motion of material objects: they were attracted or repelled, he said, by the competing forces of *philia* (love) and *neikos* (strife).

Why did Empedocles say there were four elements, rather than three or five or some other number? And why did he decide on those particular components? It's likely, scholars say, that he was influenced by what he saw around him near his seaside home: earth beneath his feet; air above him; fire in the sun, the stars, and the hearth; and water surrounding all. As well, the number four is thought to have had a special place in the mathematics of the early Greeks.

Empedocles had rationalized the *kind* of materials that objects were made of, but what would you actually *see* if you could probe matter at the smallest scale? With no microscopes to answer that question, the Greeks used reason alone to put forward an educated guess. This step comes from two thinkers who lived in the second half of the fifth century B.C. The first was Leucippus, a mysterious figure about whom we know almost nothing. Of his writings, we have one solitary fragment: "Nothing happens in vain," he said, "but everything for a reason and by necessity." The second, a follower of Leucippus, was a man named Democritus.

Democritus and the Atom:
A Big Idea about Something Small

Democritus (*c.*460–*c.*370 B.C.) hailed from Abdera, a city in Thrace, the Greek province on the northern shore of the Aegean. He may have been a pupil of Leucippus; their exact relationship is unclear. He seems to have been a jack of all trades, writing on physics, astronomy, mathematics, music, literature, and ethics— some 50 works in all. None of these survived, yet enough was written about him and his achievements to secure a place for him among the giants of early Greek thought.

Democritus, like Empedocles, asked about nature's building blocks. If you cut a log in two, and then cut it again and again ...what would be left? With a perfect knife, could you just keep cutting forever? No, reasoned Democritus, there has to be some lower limit. There must be some point beyond which matter can no longer be divided, some basic entity that cannot be cut. He called that entity *atomon*—literally, "uncuttable." Today we call them *atoms*, the basic constituents of matter. In fact, Democritus said, the process of cutting is an illusion. When we say we're cutting an object in two, what we really mean is that we're inserting a knife into the empty space between the atoms, pushing some atoms to one side and some to the other. When we come down to a single atom, such a division is no longer possible.

For Democritus, atoms were fundamental. The complexities of nature, the behavior of men and beasts—all were the result of different kinds of atoms coming together in various configurations. The most famous epigram attributed to him (second-hand, of course) underscores this belief: "By convention color, by convention sweet, by convention bitter; in reality nothing but atoms and the void."

Atoms come in an infinite number of different shapes and sizes, Democritus reasoned, though each individual atom is eternal and unchanging. He even suggested the mechanism by which atoms could attach to one another:

> The atoms have all sorts of shapes and appearances and different sizes ...some are rough, some are hook-shaped, some concave, some convex, and some have other innumerable variations....Some of them rebound in random directions, while others interlock because of the symmetry of their shapes, positions, and arrangements, and remain together. This is how compound bodies were begun.

We can think of these atoms as nature's Lego™ set: each is equipped with protrusions or holes that allow it to latch on to its neighbors. The atoms themselves may be simple, but with billions of them arranged in endless combinations, they can form objects of all shapes and sizes.

Democritus, like Thales and Empedocles, probably deserves the label "materialist"—he sought material, or physical, explanations for what he saw in nature. He viewed the natural world as a series of causes and their logical effects, not as the playground of the gods. And seeking out those links between cause and effect was his passion. He once remarked that he'd rather discover a single new cause than become King of Persia.

Yet in those days the dividing line between science and religion was, to say the least, blurry. Today, we often hear that the two disciplines have carved off separate realms of inquiry (although, as we'll see in the final chapter, in some areas they could still be seen as overlapping). Back then, however, science and religion were much less clearly defined. Empedocles included the Greek gods in his theory, saying that they, too, were composed of his four elements. He may even have equated a particular deity with each element. Democritus, however, kept the gods in the back seat; they had little to do with his atoms. Nature's building blocks, he argued, moved without any grand purpose or design.

Trying to draw a line between the science and the religion of the ancient Greeks is probably futile. More importantly, science and religion can be seen to stem from a common root, the two branches growing and evolving alongside one another, even influencing each other. And the subject of this book—the quest for a single, unified theory—may have its origins in both disciplines. John Barrow, a British physicist and author at the University of Cambridge, says the search for simplicity is probably older than science itself, dating back to the first myths and legends—the tales that kept our ancestors enthralled as they gathered around the campfire to share their stories and ideas. The search for a single theory, Barrow says, stems from an ancient desire for intellectual security—a desire to tell ourselves that we know everything:

> If you look back at the earliest myths and legends that the first cultures had about the nature of the world, [you see that] they tried to join everything together and to explain absolutely everything. Those were, in many ways, the first Theories of Everything. They didn't want to leave anything out. They didn't want to threaten their security by having something that was labeled as "unknown" or "unknowable" in the world around them. So I think there is a deep—what we might call religious—inclination to try to produce a single, coherent account of everything around us.

The Greeks took the first bold steps beyond mythology, giving us a new way of thinking about the world. But early science shared something with mythology and religion. Like those more ancient ways of describing the world, science was searching for simplicity, for a concise but complete explanation of the universe. It is a search that continues today.

After the Presocratics

The Presocratics were only the first players in this intellectual drama. Later thinkers would take Greek civilization—and Greek science—to even greater heights. Euclid (*c.*330–*c.*260 B.C.) is remembered for his massive and enormously influential treatise, *The Elements of Geometry*, which sets out hundreds of mathematical problems and their solutions in a series of explicit theorems and proofs. It served as the definitive geometry textbook for more than 2000 years. Archimedes (287–212 B.C.) was an equally great mathematician as well as a brilliant inventor and engineer; the "Archimedes screw" is still used in Egypt to draw river water for agriculture. After discovering the law of buoyancy by stepping into a full bathtub, he's said to have run naked through the streets of Syracuse, shouting "*Eureka!*" ("I have found it!"). The Greeks' mastery of mathematics may have been their greatest contribution to Western science. "Greek geometry and speculative thought," writes historian Alan Cromer, "were unique inventions, never duplicated by other cultures, even those that had engaged in some kind of mathematics for thousands of years."

Yet not every contribution was a positive one. Consider the case of Aristotle (384–322 B.C.). He was, without question, a brilliant thinker; indeed, when asked to name the greatest philosopher of all time, scholars usually point to either Aristotle or his teacher, Plato (427–347 B.C.). A tireless observer and scholar, Aristotle wrote extensively on logic, ethics, rhetoric, politics, natural history, and metaphysics. In the physical sciences he was equally prolific, studying the atmosphere, thunder and lightning, earthquakes, and mineralogy. Yet certain aspects of his approach to physics were—from a modern perspective—deeply flawed. Instead of looking for the *causes* of natural phenomena, Aristotle focused on the search for *purpose*—and, by doing so, took physics toward an intellectual dead end. Consider the question of why a heavy object falls to the ground. For Aristotle, the answer was simple—it must be striving toward the lowest possible location, the solid earth. Heavy objects, he reasoned, must fall faster than lighter ones. The "purpose" of a heavy object is to come to rest on the ground, its "natural place," and would do so more quickly than a lighter object. The idea was plausible enough, given what Aristotle knew about the world—in air, a hammer really does fall faster than a feather—but it was hardly a world-view that encouraged investigation or experiment.

Aristotle also contemplated the structure of the cosmos. He envisioned the universe as a series of concentric, crystalline spheres which carried the sun, moon, planets, and stars around

The Greek philosopher Aristotle. Though he was a brilliant thinker, his philosophy discouraged experimentation.

Alinari/Art Resource, N.Y.

the earth. (Five planets—Mercury, Venus, Mars, Jupiter, and Saturn—were known in antiquity; Uranus, Neptune, and Pluto were discovered only in modern times.) Attached to these spheres, the planets, by definition, would move in perfect circles—the circle, in turn, being the most perfect form known in geometry. The spheres, he said, must be immutable; only below the level of the moon—the lowest sphere above the earth—was change possible. And it was only in this innermost realm, Aristotle said, that matter was composed of the four traditional elements: earth, air, fire, and water. Above, in the realm of the planets and stars, there was a different kind of substance—a fifth element, or *quintessence*. Meanwhile, the earth—corruptible and imperfect—lay motionless at the center of these great spheres.

The Egyptian astronomer Claudius Ptolemy (*c.*90–168 A.D.) later absorbed Aristotle's ideas into a more complete *cosmology*— a comprehensive, mathematical model of the motions of the sun, moon, stars, and planets. He set down his ideas in a voluminous textbook that became known as the *Almagest*—derived from an Arabic phrase meaning "the Majestic" or "the Great." Founded on Aristotle's earth-centered or *geocentric* model of the heavens, the mighty *Almagest* would serve as the core of all astronomical teaching for a staggering 14 centuries.

The Legacy of the Greeks

To do justice to what they initiated ...we must regard them at least as protoscientists, standing at the gateway of that part of ancient philosophy that was called physics.

A.A. LONG

How direct is the link between ancient Greek science and modern physics? In some fairly obvious ways, the Greeks were wrong. The atom turned out to be divisible after all, and modern chemistry boasts more than 100 elements rather than the four embraced by Empedocles. And the cosmology of Aristotle and Ptolemy, as we'll see shortly, would face a profound challenge. But in many ways the Greeks were right on target. It turns out that the physical world really is explicable in terms of invisible component parts that come in just a few varieties. Compare Democritus' view—that there is "nothing but atoms and the void"—with that of quantum theory pioneer Erwin Schrödinger: "Matter is constituted of particles, separated by comparatively large distances; it is embedded in empty space." The approach of the ancient Greeks was, in fact, remarkably modern. As historian Barnes writes: "If their attempts sometimes look comic when they are compared with the elaborate structures of modern science, nonetheless the same desire informs both the ancient and the modern endeavors—the desire to explain as much as possible in terms of as little as possible."

We remember the ancient Greeks not for the conclusions they reached, which were often flawed, but for the reasoning they used to get there, which—given what they knew of the world—was sound. For the first time, people began to study the material that makes up the natural world and the forces that govern its behavior. And from the variety, confusion, and disorder that they saw, they tried to grasp the underlying principles—simple, compre-

The Egyptian astronomer Claudius Ptolemy. His textbook on astronomy, the *Almagest*, held sway for 14 centuries.

hensible, natural laws. The Greeks may not have been as sophisticated as today's physicists, and their methods were of course limited, but their goal was the same. With this first flowering of rational inquiry, writes physicist and historian Bernard Pullman,

...the principal ingredients of a scientific approach are beginning to take form: the drive to explore the universal and the essential; the belief that nature, under its complexity and astonishing diversity, hides an order that can be articulated in terms of simple elements and their interactions; the hope that, in the best of cases, a unifying reason might even preside over the extraordinary variety and endless changes of the elements of nature; and, above all, the conviction that, in this grand cosmic puzzle, only rational elements and events intervene—in other words, there is no place for supernatural mediation.

The Greeks, then, were the first to search for unification in nature—not by appealing to the gods, but by examining the world and seeking order from within the apparent chaos. They wanted to find an explanation that was at once simple and all-encompassing. In short, they were looking for a Theory of Everything. Had T-shirts rather than tunics prevailed in ancient Greece, we can imagine Thales sporting the slogan "All is Water," or Democritus with the catch-phrase "It's All in the Atoms." The physicists we'll meet in the coming chapters continued the search begun by these early thinkers. Progress, however, was not always swift; atomic theory, for example, would languish for 23 centuries. The flame of scientific inquiry would not always burn brightly—but, once lit, it would never be extinguished. The adventure had begun.

A New Vision

The Copernican Revolution

O Lord my God ... Who laid the foundations of the earth, that it should not be moved forever.

PSALM 104

It is clear that the earth does not move, and that it does not lie elsewhere than at the center.

ARISTOTLE

What the Greeks had achieved was tremendous—but it didn't last. After reaching its climax in the fifth and fourth centuries B.C. with great thinkers like Socrates, Plato, and Aristotle, the Greek world began to fall apart. The following centuries brought warfare among the Greek city-states, attacks from the Macedonians to the north, and finally conquest by the armies of Alexander the Great. With the loss of political stability came the end of prosperity; without prosperity, leisure was lost; and without leisure, the pursuit of science was lost.

Our story, then, must take a slight detour; it is, after all, hard to search for a unified theory of physics when no one is practicing physics. To connect Greek science with today's science, we must explore this disjointed path, following the winding route that science took after this first great experiment in rational inquiry came to an end.

After Alexander's soldiers came those of the growing Roman Empire—but Rome could not duplicate the atmosphere of intellectual curiosity that had flourished in Greece. The Romans were keen on law and history and produced great works of literature; they were also master artists and architects. But they had little time for pure science, especially the kind of theoretical musings that we saw in Greece just a few centuries earlier. Finally Rome, too, collapsed (or rather, its western half collapsed) as invading tribes from northern and eastern Europe stormed into Italy in the fourth and fifth centuries A.D.

One element of the Roman world, however, survived. Christianity, the religion of the late Roman Empire, offered a message of inspiration and hope for the poor, cold, and hungry of Europe. It promised salvation in the next world, but did little to nurture an interest in *this* world. "It is not necessary to probe into the nature of things, as was done by those who the Greeks call *physici*," declared the Christian theologian, Augustine of Hippo

"Hobart, this is Merlin, my science adviser."

(354–430). "It is enough for Christians to believe that the only cause of all created things ...is the goodness of the Creator, the one true God." The observation of nature was "deliberately stunted" in Christian Europe during the Middle Ages, writes historian Thomas Goldstein. After the fall of Rome, a sense of hopelessness and despair gripped the continent; all that remained was faith in an invisible and unknowable world beyond present-day suffering. The Earth and the heavens "no longer seemed a worthy object of intellectual scrutiny," writes Goldstein. There was "no room for scientific observation in this medieval, transcendental view of the world."

In the eastern provinces of what had been the Roman Empire, now the Byzantine Empire, things were very different. The Greek language was kept alive and Greek manuscripts preserved. When Arabic-speaking Muslims swept through the region in the seventh and eighth centuries, they absorbed Greek knowledge from the lands they conquered. The Muslims soon controlled much of the Mediterranean basin, including Sicily and Spain, and the entire Middle East.

For some 500 years, the Arabs were the world's curators of scientific knowledge, keeping the ideas of the Greeks alive and adding their own inventions and discoveries. They built universities, libraries, and hospitals. They also built great observatories and used astrolabes and quadrants to map the heavens. They cataloged hundreds of stars, and their legacy survives in dozens of star names such as Aldebaran, Deneb, Rigel, and Betelgeuse. They studied navigation and invented the magnetic compass. Borrowing from India, the Arabs developed the system of counting that we use today, based on what we now call "Arabic numerals" and on the place-value system. In the new system, a "1" could mean 10 or 100 or 1000, depending on where you put it and on how many zeroes (also a borrowed Indian invention) you put after it. (If you don't believe this was a great moment in the history of science, just try adding up a list of numbers using Roman numerals!) And—much to the chagrin of schoolchildren around the world—the Arabs invented algebra and trigonometry. And, perhaps above all, they translated and preserved countless Greek texts. While Western Europe plunged into the darkness of the Middle Ages, the Arab world kept the flame of learning—including science—burning bright.

Centuries passed before this storehouse of wisdom spilled into Europe. The process began in Sicilian and Spanish towns, where cultures and languages mingled and customs collided. When Christians re-conquered these lands from the Muslims, they inherited a treasure-trove of knowledge spanning 15 centuries. Ideas written in Greek and enhanced in Arabic were now translated once again—this time into Latin, the language of learning in

Christian Europe. In Spanish cities like Toledo and Cordoba, the marketplaces hummed with Arab, Jewish, and Christian traders; behind the scenes, multilingual scholars diligently copied text after text—manuscripts penned in Greek more than a millennium earlier and now enriched with the ideas of Arabia and India. Their work triggered an explosion of learning that changed the world.

The Dark Ages Weren't So Dark

In expounding Scripture, when the event described admits of no natural explanation, then and then only should we have recourse to miracles.

ANDREW OF ST. VICTOR, 12TH CENTURY

That wave of learning found its first home in the monasteries and cathedral schools that were springing up across Europe. Many were founded by clergymen of the Franciscan order, known as the "grey friars," established in 1209, or by the Dominicans, the "black friars," founded in 1215. With Christianity now the dominant cultural force across the continent, these centers of worship and prayer were also becoming centers of scholarship; by the late Middle Ages they even embraced science, at that time called *natural philosophy*. (The word "scientist" entered common usage only in the nineteenth century, though I use it here to describe those who used the methods of science to study nature.) In the wake of the Galileo affair—we'll get to him in the next chapter—we often think of the Church as being inherently hostile to science; in reality, the two usually lived in harmony. Religious leaders, in fact, relied on the work of astronomers. The date of Easter, the holiest day in the Christian calendar, could only be determined through careful observations of the heavens.

Ptolemy's view of the universe, with the earth at the center and the sun, moon, planets, and stars revolving around it. This *geocentric* picture was the accepted view throughout the Middle Ages.

The monasteries were not the only institutions to house those who could read, write, and teach. In the eleventh and twelfth centuries, Europe's first universities were founded—Bologna in 1088, Paris around 1120, Oxford about 1175. By 1400, nearly every European nation had one. Medieval universities, however, were not exactly hotbeds of intellectual dissent. Professors were not expected to show innovation or originality, only to select, preserve, and pass on traditional knowledge. In the case of natural philosophy, that meant teaching the texts of Aristotle and his followers, including Ptolemy.

As we saw in the previous chapter, Ptolemy built up a model of the cosmos based on Aristotle's earth-centered, or geocentric, view, in which the heavenly bodies are carried across the sky on crystalline spheres. The problem wasn't that Aristotle and Ptolemy ignored what they saw in the sky—in fact, their goal was to establish a model that would allow astronomers to accurately predict the future positions of the heavenly bodies. Given their knowledge of the apparent motion of those bodies, the system of crystalline spheres was a perfectly reasonable model. Their description of the cosmos was self-consistent and logical—but it was too rigid. Instead of observing whether the planets moved in perfect circles, they assumed the motion was circular and struggled to make any and all observations conform to their theory. The model of a fixed earth surrounded by perfect celestial spheres had become so entrenched, so immune to questioning, that no room was left for modifications or corrections. It went unchallenged for more than a millennium.

Christianity gradually adopted the world-view of Aristotle and Ptolemy—and it's not difficult to see why. The theory seems to be supported by "common sense"—the earth, to a casual observer, really does appear firm and immobile beneath our feet; the stars and planets do indeed seem to revolve around us. Secondly, it seems consistent with accounts in the Bible (for example Psalm 104, cited above). In another famous passage, Joshua commands the sun—not the earth—to stand still in order to prolong the hours of daylight, allowing the Israelites to finish an ongoing battle.

Dogma was plentiful in medieval Europe; there were some positions one simply didn't question. Yet it was hardly an intellectual wasteland. By the thirteenth century, a number of scholars were promoting a remarkably modern view of science. Albertus Magnus (c.1206–80), a German naturalist, was the first scholar in nearly a thousand years to make detailed studies of insects, birds, and mammals. Nature, he said, was something to be seen with one's own eyes, not just read about in books. His student, the theologian Thomas Aquinas (1227–74), gave reason and revelation equal footing in the search for truth. In England, Robert Grosseteste (c.1175–1253), Roger Bacon (c.1220–92), and John Pecham (d.1292) studied light and optics, and emphasized the value of experiment. Another Englishman, Andrew of St. Victor (late twelfth century), urged rational inquiry over blindly attributing natural phenomena to God. Perhaps the most famous was William of Ockham (c.1285–1348), a Franciscan monk who taught at Oxford and Paris (his name is sometimes spelled "Occam"). When comparing two different theories, Ockham argued, the one that makes the fewest assumptions about the known facts is the better explanation. (In his own words, "entities are not to be multiplied beyond necessity.") The argument, now known as "Ockham's Razor," still resonates with today's scientists.

Science slowly began to emerge from the shadows of medieval philosophy. But one field in particular—astronomy—would deal the final blow to the medieval European world-view. Modern science, and especially modern physics, could only see the light of day once Aristotle's crystalline spheres had been dismantled.

Copernicus: Shattering the Spheres

Some people, as soon as they find out about this book I have written on the revolutions of the universal spheres, in which I ascribe a kind of motion to the earthly globe, will clamor to have me shouted down.

NICOLAUS COPERNICUS

Nicolaus Copernicus (1473–1543) lived during an age of unprecedented exploration and discovery. He was 19 when Columbus made his first voyage to America and 24 when Vasco da Gama rounded Africa to reach the Indian Ocean. It was also an age of ideas. Thanks to the printing press, books were now cheap enough that they could be found in every corner of Europe. Those treasured texts of Aristotle and his fellow philosophers, once copied laboriously by hand, could now be mass-produced. By one estimate, six to nine million copies of more than 35,000 titles had been printed by 1500.

Copernicus was born in the Polish city of Torun. He studied in Cracow, Bologna, and Padua before returning to his homeland, where he became a canon at Frombork Cathedral. By this time, he had learned the inner workings of Ptolemy's astronomy. The more he studied the Ptolemaic model, however, the less he was satisfied by it. First of all, a simple earth-centered model of the solar system could not account for the observed motions of the planets. Among the difficulties:

- If the paths of the planets are smooth circles, why do Mars, Jupiter, and Saturn sometimes appear to "backtrack" in their paths across the sky?

- Why do the planets appear to speed up and slow down as they move along their orbits?

- If all the planets revolve around the earth in circular paths, that means they're always at the same distance from us. Why, then, do the planets—especially Venus and Mars—vary so much in brightness over the course of a year?

In order to explain these observations, the earth-centered model of Ptolemy needed substantial tweaking. Instead of moving in simple circles around the earth, each planet was also presumed to move in one or more smaller circles known as *epicycles.*

Nicolaus Copernicus. His book was the first widely publicized challenge to Ptolemy's earth-centered view.

Yerkes Observatory / University of Chicago

Ptolemy's system needed dozens of these epicycles. To Copernicus, this hodge-podge of orbits and epicycles seemed grossly inelegant. The astronomers who supported it "have been like someone attempting a portrait by assembling hands, feet, head, and other parts from different sources," he wrote. "These several bits may be well painted, but they do not fit together to make a single body. Bearing no genuine relationship to each other, such components, joined together, would compose a monster, not a man."

Copernicus offered a simpler solution: perhaps the sun, not the earth, was the center of the solar system. The idea of a sun-centered system had been suggested by a few ancient Greek astronomers; we know such a view was taught by Aristarchus of Samos (c.310–230 B.C.) some 18 centuries earlier. However, no one had worked out the details of the theory and the idea was abandoned. But Copernicus realized that a sun-centered (or *heliocentric*) system solved many of the problems that plagued the Ptolemaic model. The variations in the brightness and speed of the planets, as seen from the viewpoint of a moving earth, now suddenly make sense. Even the backtracking of the outer planets, what astronomers call *retrograde motion*, is a natural result of the sun-centered model. As the fast-moving earth passes or "laps" the slower-moving outer planets—Mars, Jupiter, and Saturn—they appear to temporarily reverse their direction as seen against the background stars.

There were, of course, objections to the Copernican model, many of them based on "common sense." If the earth really moves, why does a ball tossed into the air come down at the same spot it was thrown from? Why aren't birds swept backwards, against the direction of the earth's motion? Indeed, Aristotle himself had ridiculed the idea of a moving earth for just such reasons. (These objections, incidentally, can be raised against the earth's daily rotation, as well as its yearly journey around the sun.) Copernicus, however, had an answer: the earth's atmosphere must be carried along with it, so that no such motion is seen. (A complete explanation requires the concept of *inertia*, the tendency of moving objects to remain in motion and of stationary objects to remain at rest. The term, as we use it today, made its appearance only with the work of Kepler and Galileo, and would become a cornerstone of Newton's mechanics.)

And there were more objections: if the earth really moved around the sun, the stars should appear to shift in position over the course of a year; in astronomical jargon, they should display *parallax*. (As an analogy, imagine walking through the woods. As long as you're moving, the nearby trees appear to shift their position relative to the more distant trees behind them.) Yet no such shift was seen in the positions of the stars. And, if the earth moves in a vast orbit, then it must be nearer to certain stars at certain times of the year; therefore, the stars should vary in brightness over the course of the seasons. The answer, Copernicus reasoned, is that the stars must be very far away compared to the size of our solar system.

This view of a large, possibly infinite universe was a radical departure from the popular medieval view, in which the heavens were thought to be just out of reach, hovering just a little higher than the highest mountains. True, astronomers who actually measured the motions of heavenly bodies understood that the cosmos was indeed vast—but now it became clear that the stars were so distant as to defy the imagination. Yet for Copernicus, this new, larger, universe—with its more coherent cosmological picture—was easier to swallow than Ptolemy's countless epicycles. "I think it is a lot easier to accept this [larger universe] than to drive ourselves to distraction multiplying spheres almost ad infinitum," he wrote, "as has been the compulsion of those who would detain the earth in the center of the universe."

Several myths have grown up around the Copernican Revolution. One is that his heliocentric system solved all of the problems in astronomy in one swoop. In fact, it did not: in order to precisely match the observed motions of the sun, moon, and planets, his system also needed epicycles. And calculating planetary positions using the Copernican model could be even more cumbersome than with the old Ptolemaic system. That's because Copernicus's theory contained a vital error. Like the ancients, he was committed to the idea that circles and spheres were the embodiment of perfection, and assumed that the orbits of the planets were perfectly circular. It would be another 70 years before Johannes Kepler would deduce their true shape.

Another myth is that religious leaders opposed a sun-centered system because it deprived humanity of a "special place" in the universe. In fact, the center of Aristotle's universe was reserved for hell—not exactly a distinction to envy. The idea of a moving earth was more upsetting, although as long as the Copernican model was used as a "mathematical tool" (a theoretical device to aid in calculations), few had objections. There was, however, reluctance to accept the vast, possibly infinite cosmos that the Copernican model suggested. (By one sixteenth-century estimate, the Copernican universe would have to be 400,000 times larger than that of Ptolemy.) Religious believers "had to adopt this new idea," says Harvard astronomer and historian Owen Gingerich. "They had to adapt to a moving earth, and to a larger universe—so that heaven was not so tidily arranged, just beyond the sphere of the planets, not all that far away. Getting used to a much larger scale ...has been, I think, very troubling to the human-sized view of the cosmos that was inherent within the Bible."

Copernicus, not surprisingly, was reluctant to publish his theory. Only in his old age did he finally allow his great work, *De Revolutionibus Orbium Coelestium* (*On the Revolutions of the Celestial Spheres*), to be printed; a finished copy was presented to him on his deathbed. Unknown to Copernicus, his publisher had added a preface—essentially a disclaimer—saying that the heliocentric system was just a theoretical model.

But Copernicus knew his model was more than just an abstraction. Contemplating its sense of balance and beauty, he wrote that "we discover in this orderly arrangement the marvelous symmetry of the universe, and a firm harmonious connection between the motion and the size of the spheres...." The sheer elegance of the heliocentric model—the *idea*, not the details—was enough to convince him that it was a true description of nature.

Tycho Brahe: The Great Observer

The next challenge to medieval cosmology came from one of the most unusual heroes of Renaissance astronomy, Tycho Brahe of Denmark (1546–1601). Tycho's contributions to science have often taken a back seat to the peculiar details of his personal life. The story of his nose is perhaps the most dramatic. As a 20-year-old student, he found himself in a duel with one of his classmates. His opponent's sword sliced through the bridge of his nose, removing a sizable chunk. For the rest of his life, Tycho wore a metal prosthesis to hide the unsightly gash; his companions sniggered each time he applied a dab of ointment from a jar he kept by his side. One of his enemies, Nicolaus Ursus, said it was little surprise Tycho was able to see multiple stars "through the triple holes in [his] nose." But Tycho was not a man to let a physical flaw get in the way of his ambitions. In time, he became the greatest astronomer in Europe.

Tycho—like Galileo, history remembers him by his first name—was born in the Danish province of Scania (today part of southern Sweden) three years after the publication of Copernicus's revolutionary book. As a young man, Tycho witnessed a series of events in the heavens that sparked a fascination with the night sky. The first was the solar eclipse of 1560. Tycho, a wide-eyed teenager, was amazed that astronomers could predict such events months and even years in advance. Later, while studying in Germany, he witnessed a close pairing of Jupiter and Saturn (astronomers call it a *conjunction*), an eye-catching celestial gathering that occurs about once every 20 years. But Tycho noticed that the published tables, whether based on Ptolemy's ancient system or the newer Copernican model, were grossly inaccurate; the time given for the closest approach of the two planets was off by several days. Suddenly, Tycho knew his true calling: he would devote the rest of his life to making the most accurate possible observations of the heavenly bodies.

An even more spectacular event lit up the skies in November, 1572, when a bright "new star" appeared in the constellation Cassiopeia. (Today we would call it a *supernova*, the explosion that takes place when a massive star exhausts its nuclear fuel supply.) Tycho described the event in his book, *De Stella Nova* (*On the New Star*):

> I noticed that a new and unusual star, surpassing all the other stars in brilliancy, was shining almost directly above my head. And since I had almost from boyhood known all the stars of the heavens perfectly ...it was quite evident to me that there had never before been any star at

that place in the sky, even the smallest, to say nothing of a star so conspicuously bright as this.

But the established cosmology of the day forbade any changes in the starry heavens. Any new object, such as a comet or Tycho's new star, was assumed to be an atmospheric phenomenon, located below the sphere of the moon. Yet if the new star were this close to the earth, Tycho reasoned, it should display parallax—that is, its position relative to the background stars should change over the course of the night as the earth rotated. But his new star showed no discernable parallax. "I conclude," he wrote, "that this star is not some kind of comet or fiery meteor ...but that it is a star shining in the firmament itself—one that has never previously been seen before our time, in any age since the beginning of the world."

Tycho's observation of the new star dealt a shattering blow to the established cosmological order. Those who wanted to cling to the universe of Aristotle and Ptolemy might have dismissed the Copernican model of the solar system as a mathematical convenience, but there was no escape from Tycho's supernova, seen by skywatchers across Europe and now proven to lie in the supposedly immutable heavens.

The observation had another crucial effect: it made Tycho famous. So famous, in fact, that in 1576 the Danish king, Frederick II, made him a remarkably generous offer: an island of his own, in the channel between present-day Denmark and Sweden, where Tycho could pursue his dream of mapping the heavens. "If you want to settle down on the island," the king wrote, "I would be glad to give it to you as a fief. There you can live peacefully and carry out the studies that interest you, without anyone disturbing you....I will sail over to the island from time to time and see your work in astronomy and chemistry, and gladly support your investigations." It was an offer Tycho couldn't refuse.

Beginning in 1576, Tycho transformed the small island of Ven into Europe's foremost center of astronomical learning. Within a few months, Tycho and his assistants were observing the sun, moon, planets, comets, and stars. They invented new tools for astronomy and map-making, and used their own printing press to share their findings with the world. Learned young men from across the continent descended on Tycho's laboratory, known as Uraniborg ("heavenly castle"), eager for a chance to work with the famous observer.

Tycho spent 21 years on Ven, until a series of personal and political difficulties forced him to leave Denmark. In 1599 he took a position in Prague under the patronage of Emperor Rudolf II. Tycho's career was winding down, though he was still keenly interested in matters of astronomy. He proposed a new model of the solar system, a compromise between that of Ptolemy and Copernicus. In Tycho's model, the planets revolved around the sun, but the sun and moon revolved around the earth. The "common sense" stability of earth, the absence of stellar parallax, and the ancient prohibition against earthly motion made it impossible for Tycho to embrace Copernicus's larger, sun-centered system.

By this time, Tycho had heard about a young German scientist, Johannes Kepler, who was making a name for himself in mathematics and astronomy. Impressed with his colleague's skill, he invited Kepler to join him in Prague. Tycho, unfortunately, had only a year left to live, so their time together was brief.

The peculiar and tragic story of Tycho's death, like the story of his nose, often overshadows his contributions to science. In the autumn of 1601, Tycho was invited to a dinner party hosted by one of Prague's most important noblemen. Part-way through the dinner, Tycho realized he had to go to the bathroom. Rather than excuse himself from the table and risk offending his host, he held it in. Kepler, writing on the final page of Tycho's log book, described what happened: "Holding his urine longer than he was accustomed to doing, Brahe remained seated. Although he drank a bit overgenerously and felt pressure on his bladder, he had less concern for the state of his health than for etiquette." By the time the party was over, it was too late; Tycho died from complications 11 days later.

Johannes Kepler:
The Harmony of the Heavens

On a hilltop overlooking Prague's medieval castle stands a larger-than-life bronze sculpture of two scientists. On the left is Tycho Brahe, clutching a giant sextant; by his side is Johannes Kepler, holding a scroll or a parchment—containing some abstruse calculations, no doubt. (Oddly, the sculptor has depicted Kepler, the theoretician, with a skyward gaze, rather than Tycho.) What the sculpture doesn't show is the often prickly relationship between the two men—Tycho, the wealthy, plump, arrogant Dane with the extravagant clothes and the metal nose; Kepler, the weaker, withdrawn, mystical German. Indeed, Kepler and Tycho may be the ultimate "odd couple" in the history of science.

Johannes Kepler (1571–1630) studied at a number of German and Italian universities before returning to his homeland to enroll in the theological seminary at the University of Tubingen. Along with theology, Kepler studied philosophy, mathematics, and astronomy, and was lucky enough to have one of Germany's leading astronomers, Michael Mastlin, as one of his teachers. Publicly, Mastlin taught the established Ptolemaic astronomy; privately he

(Author photo)

The Danish astronomer Tycho Brahe (left) and the German mathematician Johannes Kepler, shown here in a Prague statue.

instilled in Kepler his awe and admiration for the Copernican system.

Kepler was well on the way to becoming a Lutheran minister when he was asked to take on a teaching job at a provincial high school in Graz, Austria. At first, he was disappointed to be called away from his theological studies—but soon, immersed in the world of astronomy and mathematics, he had a change of heart. "I had the intention of becoming a theologian," he wrote to his teacher, Mastlin. "For a long time I was restless: but now I see how God is, by my endeavors, also glorified in astronomy."

We remember Kepler as a scientist, but he lived at a time when science, pseudoscience, and magic were all very much intertwined. Kepler was struck by the special properties of certain numbers, and devoted countless hours to developing a model of the heavens that would conform to his notions of mathematical beauty. He was fascinated by parallels between mathematics and music, and longed to translate the positions and motions of the heavenly bodies into a musical score. He wanted a solar system that would please the ear and the mind as well as the eye. Kepler was particularly fascinated by the five "regular solids" of geometry—the tetrahedron, cube, octahedron, dodecahedron, and icosahedron (pictured below). In what he must have thought was a flash of pure genius, he speculated that the orbits of the five known planets had the same proportions as these five perfect geometrical figures. He was wrong—but, as we'll see shortly, his dedication to mathematics and his sheer inventiveness eventually paid off.

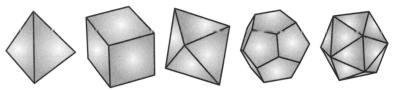

Kepler was inspired by what he regarded as the perfection of the five "regular solids" of geometry.

Kepler's scientific writings were coupled with endless speculations on metaphysics, history, and religion. As historian I. Bernard Cohen put it, "We could easily assemble a whole volume of his writings that would show how unscientific his thinking and his science were." Kepler was also a practicing astrologer, casting horoscopes for the German nobility. It's doubtful, however, that he truly believed the stars influenced our lives. He once referred to astrology as the "foolish little daughter of the respectable, reasonable mother astronomy."

Proof that the superstitions of the medieval world had not yet given way can be seen in the misfortune of Kepler's mother. When rumors spread that she was concocting mind-altering libations in her kitchen, the town authorities charged her with witchcraft. Kepler wrote numerous letters to protest her innocence, but it was not enough: she was imprisoned in 1620. Kepler eventually left his family and travelled to join his mother, spending nearly a year by her side and probably saving her from torture and execution. Only after pleading on her knees that she would rather die than confess to something she wasn't guilty of was she finally released.

We remember Kepler, however, for the fruits of his collaboration (using the term loosely) with Tycho Brahe. To his teacher, Mastlin, he complained of Tycho's "instability of character," though he added that the Dane is "a man of great benevolence." To his wife, he sounded a note of resignation: "God has united me with Tycho by an inexorable fate."

We know the two men couldn't agree on the structure of the solar system: Kepler was a staunch Copernican; Tycho, to the end, rejected the heliocentric model. While Kepler was a brilliant theoretician, he was not a particularly skilled observer—that was Tycho's department. In fact, during his years on Ven, Tycho had already accumulated all the data Kepler could ever want, but the stubborn Dane was reluctant to hand over his observations. Even on his deathbed, Tycho is said to have implored Kepler not to use his data to support the Copernican system. After Tycho's death, Kepler was forced to negotiate with his heirs in order to gain access to those volumes of astronomical data.

Finally, he got what he needed. Putting his powers of mathematical deduction to work, Kepler looked for a pattern—a *simple* pattern—in Tycho's observations of the planets. Eventually, he found that simplicity: he discovered that the orbit of each planet was not a circle, as everyone had supposed; rather, it was an *ellipse* (a slightly "flattened" circle). With Copernicus's circles replaced by Kepler's ellipses, everything fell neatly into place; no more awkward epicycles were needed.

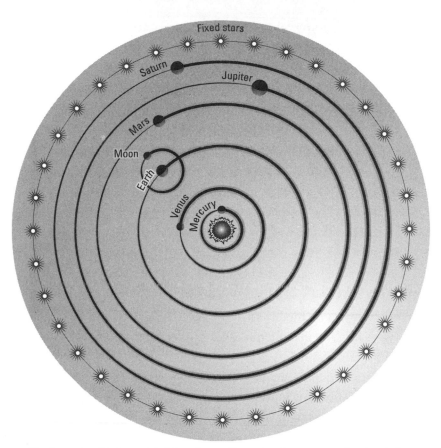

The Copernicus/Kepler model of the universe. The system is now sun-centered, or *helio-centric*: only the moon orbits the earth, with the earth and the other planets revolving around the sun. The orbits are actually ellipses though at this scale they appear circular. (The sizes of the orbits are not to scale.)

Kepler soon discovered two more laws of planetary motion: a link between a planet's speed and its position along its orbit, and a link between a planet's orbital period (the time that it takes to complete one revolution) and its average distance from the sun. It was the work of a mathematical master. No wonder the philosopher Immanuel Kant called him "the most acute thinker ever born." Or, as Albert Einstein would write three centuries later: "How much inventive power, how much tireless, obstinate work was necessary to reveal these laws, and to establish their certainty with great precision ...can hardly be evaluated by anyone."

Kepler did not have all the answers. For one thing, he believed that a planet's trajectory was governed by *magnetism*, the force that makes a compass needle point north. (Newton, as we'll see, would later correct the error and attribute planetary motion to *gravity*, the force that governs falling bodies. We must remember, though, that neither force was clearly understood in Kepler's day.) Yet his discovery of elliptical orbits was a breakthrough. Kepler, captivated by the simplicity and elegance of his model, knew the heliocentric system was more than just a mathematical convenience. It was, he was certain, the true description of the heavens. "Others may take any attitude they please," he wrote. "[But] I take it as my duty and special task to defend it before the world ...with all the powers of my brain; for I have recognized it in my own mind as true and in contemplating it, I am filled with unbelievable delight at its beauty."

Kepler died in 1630, as the devastation of the Thirty Years' War was tearing through Europe. The war obliterated all traces of his tomb. But his reputation—resting on his three laws of planetary motion and his perfection of the Copernican system—was secure. Thanks to Kepler, the solar system was a much less mysterious place.

Like the early Greek philosophers, Copernicus and Kepler tried to find the simplest description of the natural world, focusing in particular on the starry heavens. And, like the Greeks, they were drawn to the simplicity and beauty of certain ideas. For Copernicus, it was the idea of the sun-centered system itself; for Kepler, its was the expression of that idea through mathematics, especially the elegance he found in those elliptical orbits. And

Kepler deserves an additional honor. Of all the ideas we've met so far, his was the first that we still label as essentially "correct." Sure, there would be refinements—later, we'll hear about Einstein's contribution, and even *that* is likely not the final word. (Indeed, as we'll see in the final chapter, one can argue that scientific theories are *never* final.) But Kepler's laws are, for nearly every purpose, dead on; for that reason, they are still taught to undergraduate students today.

We don't need to speculate about the T-shirt that Copernicus and Kepler would have endorsed, because you can buy it in any large museum gift shop. It's that familiar drawing of the solar system, with the sun resting firmly at the center, its family of planets—blue-green Earth among them—orbiting around. (Since Tycho rejected the sun-centered model, I suppose he wouldn't wear the shirt—but I can imagine him taking command of the shop. "The heliocentric model—ah yes, they're on the rack over there," he'd tell customers. "But they'd never have figured it out without my observations.") In Kepler's case, we might want to enrich the T-shirt by adding the equations that describe his three laws of planetary motion; they could go on the back.

And yet, the evolution that we saw in this chapter was, fundamentally, the evolution of how we *picture* the universe. The equations support that vision, but in this case, a picture is truly worth a thousand words. After a lull of more than a millennium, science—and the search for simplicity—was back on track.

Heaven and Earth

Galileo, Newton, and the Birth of Modern Science

*O telescope, instrument of much knowledge, more precious
than any scepter!*

JOHANNES KEPLER

*There are more things in Heaven and Earth, Horatio, than
are dreamt of in your philosophy.*

SHAKESPEARE, *HAMLET*

The final decades of the sixteenth century and the first years of the seventeenth were a time of sweeping social and cultural change across Europe, and a period of tremendous creativity in science, art, and literature. English theater flourished under Elizabeth I, while the English language was enriched by the plays and poems of William Shakespeare and, in the reign of King James, by a new translation of the Bible. Miguel de Cervantes perfected a new literary form with his novel, *Don Quixote*. El Greco painted stark, dramatic landscapes while Caravaggio and Rubens brought a new exuberance to the canvas.

The scientific world was also in the midst of a great transformation. William Harvey revolutionized medicine by discovering how blood circulates throughout the body. Gerardus Mercator had just produced his comprehensive *Atlas*. Mathematics, too, was coming of age: the logarithm was invented and decimal notation was becoming widespread. And René Descartes—known for his philosophy ("I think, therefore I am") but also a brilliant mathematician—would soon lay down new foundations for algebra and geometry. And towering above the rest was the Italian mathematician and astronomer, Galileo.

This period falls in the middle of what we now call the *Scientific Revolution*—the set of discoveries stretching, roughly, from Copernicus to Newton. The term was coined by historians only in the early twentieth century, with precise dates open to debate. But the impact of those discoveries is beyond doubt: they established a new, rational view of the world based on empirical science and mathematical deduction. It was the greatest flood of new ideas the world had ever seen.

New ideas, however, often meet with resistance, and the provocative new view of the universe developed by Copernicus and Kepler—soon to be confirmed by Galileo—is the prime example. The "Galileo affair"—a clash of opinions that led to the astronomer's trial and imprisonment—remains the ultimate symbol of that resistance.

Galileo: Showman of Science

As a child, Galileo Galilei (1564–1642) dreamed of the priesthood; his father, a talented but cash-strapped musician, imagined his son becoming a doctor. He enrolled at the university in Pisa, the city of his birth, where he earned a reputation for arguing with his professors and fellow students alike. He studied the texts of the ancient natural philosophers, and took classes in Greek, Latin, and Hebrew. Mathematics was also taught, but primarily as an aid to understanding the other disciplines. Some of the ideas of the classical authors struck Galileo as ridiculous—for example, Aristotle's notion that heavy objects fall faster than lighter ones. (The thought is said to have come to him as he watched a hailstorm. If Aristotle were right, Galileo reasoned, it would mean that lighter hailstones must always start falling before heavier ones, or else from a greater height, so that they could all be seen to strike the ground together.) Another inspiration may have struck as he watched the gentle swaying of an oil lamp in the cathedral. This observation marked the beginning of Galileo's interest in pendulums, and his ideas would eventually allow clockmakers to construct accurate timepieces. Galileo left the university after four years, taking with him a fascination for physics—but no degree.

While struggling to get a teaching position at one of Italy's universities, Galileo spent several years in Florence and Siena tutoring students in math. Finally, he landed a post back at Pisa. His argumentative nature was by now even more on display; he frequently launched into public attacks on the ideas of the older professors. His three-year contract was not renewed. Intellectually, though, these were fruitful years. He became interested in astronomy and wrote commentaries on Ptolemy and Copernicus. He developed a simple thermometer and began to make detailed studies of mechanics. His most original ideas, however, related to the motion of falling bodies. He had already rejected Aristotle's theory that a body's rate of fall is proportional to its weight. Perhaps, he reasoned, the speed was *independent* of weight—in other words, all bodies, light or heavy, fall at the same speed (in the absence of air resistance). For Galileo, speculation was pointless; the way to know the truth was to test it experimentally.

The Italian astronomer and mathematician Galileo Galilei placed physics on a firm mathematical foundation.

According to one account, written by one of Galileo's students a dozen years after the scientist's death, he carried out the decisive experiment from atop the cathedral's bell tower—the famous Leaning Tower of Pisa. (The tower, built between 1173 and 1350, had begun to list even before its completion.) Galileo chose objects of roughly the same shape, so the story goes, in order to minimize the effects of air resistance. He climbed the tower carrying balls of lead and ebony, perhaps, or of copper and wood. We can imagine a crowd of students and professors looking on—many of them skeptical, no doubt—as the balls plunge 54 meters to the ground. The crowd hears a single "thud" as the falling weights arrive almost in unison. And if air resistance were removed completely, their arrival would indeed be simultaneous—as astronaut David R. Scott demonstrated in 1971 by dropping a hammer and a feather on the surface of the moon during the Apollo 15 mission.

Whether the grand Pisa demonstration actually occurred has been endlessly debated; Galileo's writings say only that he used a "high tower" to test his theories. Skeptics seem to be in the majority, saying that if the Pisa experiment had really happened, we'd have more than just the one belated, second-hand account of it. Historian I. Bernard Cohen says the tale is "undoubtedly false." Had it truly occurred, muses physicist Leon Lederman, it would surely have been "a media happening, the first great scientific publicity stunt." Biographer Stillman Drake says it *could* have happened, but missing documents prevent us from drawing a conclusion one way or the other.

There are at least two good reasons not to lose any sleep over the veracity of the Tower of Pisa story. The first is that several other scientists had performed similar falling-body experiments well before Galileo; indeed, a Byzantine scholar named John the Grammarian may have done it back in the seventh century. The second reason is that Galileo found an even better way to test the laws that govern falling bodies. A ball dropped from a tower requires mere seconds to reach the ground, making measurements difficult; instead, he used an *inclined plane* to slow everything down. The principles are still the same, but the times are greater and therefore easier to measure. (And we know he *did* carry out these experiments.) What he found was a precise mathematical law governing such motion. The distance a falling (or rolling) body covers is proportional to the square of the elapsed time. He had discovered that a falling body undergoes *uniform acceleration.* That concept, foreign to Aristotle, was now established. And, if anyone doubted it, they could easily repeat the experiments for themselves.

In 1592, Galileo was awarded the chair of mathematics at the University of Padua; he remained there for 18 years, the happiest and most productive of his life. He continued his study of the pendulum and wrote treatises on fortification and military engineering. He was also becoming a devout Copernican—a shift that was reflected in his lectures on the "new star" of 1604. (By sheer luck, this supernova, also studied by Kepler, lit up the skies over Europe a mere three decades after the appearance of Tycho's star in 1572. No supernovas within our galaxy have been recorded since.)

Galileo's New World

A host of other stars are perceived through the telescope which escape the naked eye; these are so numerous as almost to surpass belief.

GALILEO

In 1609, Galileo heard rumors of a remarkable Dutch invention: a tube with two glass lenses that made distant objects appear closer. The invention of the telescope is usually credited to Hans Lippershey (*c.*1570–*c.*1619), a Dutchman who chanced upon the correct arrangement of lenses the previous year. Galileo quickly developed his own instrument—by some accounts, going from theory to working telescope in just 24 hours—and producing, as it turned out, a more powerful instrument than any that had been built before.

The telescope's value as a military tool was obvious, and that's how Galileo first showed off its capability. On August 21, 1609, he led a procession of Venetian dignitaries up the stairs of the campanile (bell tower) overlooking St. Mark's Square. Aiming the instrument toward the city's harbor, Galileo invited the assembled noblemen to look through the device. They were suitably impressed: it's said that they could identify approaching ships a full two hours before they entered the port. Officials at Padua doubled Galileo's salary and offered to extend his appointment for life. He didn't take up the offer, though, instead using it as a bargaining chip to land a position in his native province, where he became chief philosopher and mathematician to the Grand Duke of Tuscany.

While Venetian leaders wanted to spy on potentially hostile ships, Galileo set his sights higher. For millennia, human beings had gazed in wonder at the night sky; now, from a quiet garden in Tuscany, Galileo became the first to take a deeper look at the panorama of the heavens. What he saw through his telescope during those crisp winter nights of 1609 would change the world forever.

Galileo observed that the planet Venus goes through phases like our own moon, suggesting that Venus in fact orbits the sun. At the same time, he discovered four "stars" that seemed to hug the planet Jupiter, shifting their position from night to night but never straying from its side; they were, he concluded, moons that revolve about Jupiter. Both of these discoveries contradicted the notion of an earth-centered universe. (I remember the thrill, as a teenager, of setting up a small telescope and catching my first glimpse of Jupiter's moons. Of course, I knew from books and photographs what to expect; Galileo did not. One can only imagine his sheer awe at seeing those bodies for the first time *ever*.) Galileo had already warmed to the Copernican model a dozen years earlier; the images glimmering in his eyepiece now cemented his belief. And the discoveries kept mounting: he saw dark spots on the sun, as well as mountains and craters on the moon—all of which contradicted Aristotle's view of the sun and the moon as "perfect" bodies.

In 1610, Galileo published his observations in a book called *Siderius Nuncius* (*The Starry Messenger*). The treatise was an instant success; soon the name of this Italian professor was on the lips of educated men across Europe and beyond. Within five years, it was being discussed as far away as China. According to biographer James Reston, Galileo's telescopic discoveries quickly made him "the most famous man in the world."

Many scholars were impressed, even delighted, by the wondrous new universe Galileo had opened up. On the home front, however, trouble was stirring. A number of critics, clinging to the world of Aristotle and Ptolemy, simply refused to acknowledge his discoveries. Some of these learned men, he wrote to Kepler, never bothered to look through the telescope, even though he "offered them a thousand times." With great sadness, he suggested that they had simply "shut their eyes towards the light of truth."

The story of how Galileo became mired in a dispute with the Church has been told many times, but the conflict is still often seen in overly simplistic terms. The Galileo affair was not just a clash between science and religion, nor was it merely a conflict between supporters of two opposing models of the heavens. It was, rather, a murky mix of philosophy, theology, and—perhaps above all—politics. And it percolated slowly, from the time that Galileo first peered through the telescope in 1609 to his trial by the Roman Catholic Inquisition nearly a quarter-century later. We will re-examine his case to dispel some of the myths that have grown up around it—and also because the conflict, more than any other episode in the history of science, highlights the resistance that sometimes greets new ideas.

Galileo and the Church

At first, Rome appeared enthusiastic about Galileo's discoveries; Church officials had seemed pleased enough when he brought his telescope to the Vatican for a demonstration in 1611. But the professors at the Italian universities, who earned their living teaching the natural philosophy of Aristotle, joined forces against him. A young Dominican named Thomas Caccini, meanwhile, tried to make a name for himself by denouncing Galileo from the pulpit. Hoping to avert a crisis, Galileo paid another visit to Rome, urging the Church to adopt a flexible attitude toward questions of physics and astronomy. It didn't work. Instead, Galileo was told he could neither "hold nor defend" the Copernican opinion. For a few years, Galileo obeyed, working quietly on other matters in his house near Florence.

When Maffeo Barberini became Pope, taking the title Urban VIII, Galileo was thrilled. A fellow Tuscan, Barberini had been a friend of Galileo's and was known for his progressive attitude toward science and philosophy. The Pope's nephew, Francesco, was also a close friend of the scientist. In that same year, Galileo dedicated a treatise on the new scientific method, titled *The Assayer*, to the new head of the Catholic Church. The Pope, in turn, is said to have been enchanted by Galileo's wit and by a political outlook that both men seemed to share; it's said the Pontiff had *The Assayer* read to him over dinner. The Pope told Galileo he could go ahead and write about the system of the world—both the Ptolemaic and the Copernican—as long as he gave equal weight to both arguments.

And so Galileo began writing his greatest work, *A Dialogue Concerning the Two Chief World Systems—Ptolemaic and Copernican*. The book, published in 1632, was hailed across Europe as a masterpiece of literature and philosophy. By this point, however, it was not so much *what he said* but rather *how he said it* that was causing trouble for Galileo. While the *Starry Messenger* had been written in Latin, his *Dialogue* was a flowing, easily-read work penned in Italian, the everyday language of his countrymen. He presented his argument in the form of a dialogue between three characters: Salviati, Sagredo, and—representing the old Ptolemaic view—the dim-witted member of the trio, Simplicio. (The name was supposedly a translation of Simplicius, a Greek philosopher who wrote commentaries on Aristotle, but the word is also Italian for "simpleton.") Galileo had paid only lip service to the Pope's request for impartiality; the book was a thinly veiled endorsement of the Copernican system. The final straw may have come when Galileo's enemies convinced the Pope that Simplicio was a caricature of the Pontiff himself. The Pope took it as a personal insult. In a rage, he ordered the *Dialogue* placed on the Index of Prohibited Books, and Galileo was called before the Holy Office of the Inquisition.

Galileo, like Copernicus and Kepler, was a man of deep faith. Yet this was a difficult time for the Catholic Church. With the Protestant Reformation gaining ground in northern Europe, the last thing the Vatican needed was an ideological challenge on its home turf. It was just such a challenge that got Giordano Bruno (1548–1600) into trouble a generation earlier. Bruno, an Italian philosopher and mystic, had also supported the Copernican system, and—perhaps even more offensive to the Church—he argued that the earth was merely one of many populated worlds, perhaps an infinite number, in the starry heavens. Bruno was burned at the stake in 1600.

With Galileo, Rome was deeply troubled by what seemed like another affront to its authority as the sole interpreter of Scripture. Consider that famous passage from the Book of Joshua:

> Then spake Joshua to the Lord in the day when the Lord delivered up
> the Amorites before the children of Israel, and he said in the sight of

Israel, Sun, stand thou still upon Gibeon; and thou, Moon, in the valley of Ajalon.

And the sun stood still, and the moon stayed, until the people had avenged themselves upon their enemies....So the sun stood in the midst of heaven, and hasted not to go down about a whole day.

In Galileo's eyes, Joshua's feat was not meant to be tied to a literal description of celestial mechanics. "I think ...it is very pious to say and prudent to affirm that the holy Bible can never speak untruth—whenever its true meaning is understood," he wrote. "But I believe nobody will deny that it is very often abstruse, and may say things which are quite different from what the bare words signify." If one takes every biblical passage literally, he cautioned, "one might fall into error." The passage from Joshua, said Galileo, accords perfectly with the Copernican description—but the author of the passage, trying to keep things simple for the shepherds and farmers who would hear the tale, told the story as if we lived in an earth-centered universe. As we might put it today, they "dumbed it down for the masses":

...since [Joshua's] words were to be heard by people who very likely knew nothing of any celestial motions beyond the great general movement from east to west, he stooped to their capacity and spoke according to their understanding, as he had no intention of teaching them the arrangement of the spheres, but merely of having them perceive the greatness of the miracle.

Like the Church authorities, Galileo believed that Scripture could never be in error—but its interpretation, he argued, could indeed be wrong. As Galileo once said, the Bible tells us how to go to heaven, not how the heavens go (a line that he borrowed from Caesar Cardinal Baronius, the Vatican librarian). "As far as the theologians were concerned, the Copernican system was not really the issue," says Harvard astronomer and historian Owen Gingerich. "The battleground was the method itself, the route to sure knowledge of the world, the question of whether the Book of Nature could in any way rival the inerrant Book of Scripture as an avenue to truth."

The Copernican system was never actually declared heretical, though in 1620 the Church ordered all printed copies of the astronomer's great work, *de Revolutionibus*, to carry a series of corrections that stressed the hypothetical nature of the argument. (In practice, says Gingerich, it was only the Italian copies that ended up being heavily censored, while those in France and Spain—also Catholic countries—were left intact. "Apparently this was being seen as a local Italian imbroglio," he says.) And now, with Galileo having drawn the ire of the Pope, his *Dialogue* was also deemed unwelcome. Initially cleared by the Church censors, the book was banned and existing copies—that is, the ones within the Vatican's reach—were confiscated.

Crime and Punishment

Galileo, now nearly 70, was charged with "vehement suspicion of heresy" and brought to trial in Rome in 1633. He had been hoping for some sort of plea bargain, in which he could admit to having acted against Papal orders and simply agree not to teach or show further support for the Copernican system. But there was no such deal. Under threat of torture—he was shown the instruments that had been prepared for him should he refuse to co-operate—he was forced to recant his belief in the Copernican theory. According to legend, as he was led away he muttered *eppur si muove* ("and yet it moves"); while he may indeed have been thinking as much, he would certainly not have dared say those words out loud.

The conflict, historians agree, was mired in politics and the struggle for power. Galileo "was sentenced rather severely for disobeying orders, for rocking the boat when Rome was trying to have a unified view, theologically, against the Protestants north of the Alps," says historian Gingerich. "The Galileo affair is heavily fraught with particular personalities that were playing out the battle, rather than lofty issues of whether humankind was at the center of the cosmos or not." Biographer Stillman Drake has called it a fight over the "right of a scientist to teach and defend his scientific beliefs." James Reston describes it as a "question of personality, not principle."

Galileo's approach to science rankled members of the clergy and academia alike. In his view, the study of nature should not be restricted to the lofty world of the Church and the universities; rather, it should be a practice open to all, with any well-trained mind potentially able to contribute. His support of the Copernican system, says historian Carlo Rubbia, was presented "as a conclusion which the reader should reach by himself, by understanding the way in which the scientific method proceeds, a scientific method which should be applied to all walks of life. It is probably the last point which frightened some people, far more than the heliocentric theory." Perhaps Galileo's ultimate crime was to challenge the notion that only theologians had access to ultimate truths. For centuries, ordinary citizens turned to the Church for insight into the nature of the universe; now this pushy Italian was saying that anyone could observe the heavens and deduce the truth for themselves.

Condemned by the Inquisition, Galileo spent the final eight years of his life under house arrest in his villa near Florence. Deflated but not defeated, he continued his scientific inquiries with more vigor than most men his age could muster. He also received a number of visitors, including two well-known Englishmen. The poet John Milton was outraged at the way Rome had treated the scientist, and would later embrace Copernican astronomy in his epic poem, *Paradise Lost*. Philosopher Thomas Hobbes dropped by as well, and informed Galileo that his *Dialogue*—so troubling to the Italian authorities—had now been printed in England. By 1637 Galileo completed another groundbreaking work, this time on mechanics—*The Dialogues Concerning Two New Sciences.* It's said that he asked permission to leave his villa only once—when he smuggled out a copy of his new book. His friends saw to its publication, not in Italy but in the Dutch city of Leiden. Galileo died in his villa during the winter of 1642.

Only slowly—very slowly—did the Vatican acknowledge it was wrong to condemn Galileo. His *Dialogue Concerning the Two Chief World Systems* was dropped from the Index of Prohibited Books only in 1835; the Index itself was abolished in 1966. In 1992, John Paul II concluded a 13-year investigation into the Galileo affair, admitting that the condemnation was the result of "tragic mutual incomprehension." The move was long overdue. The Church's treatment of Galileo is "the central embarrassment of its modern history," writes biographer James Reston. To this day, "the face of Galileo haunts the Holy Roman Church."

Galileo, literally and figuratively, had brought the heavens down to earth. By using the telescope to probe the sun, moon, and planets, he showed that these bodies—once thought to lie in their own perfect domain—could be studied by the same instruments and techniques used here on earth. After Galileo, there was little need to invoke ideas like Aristotle's "quintessence" in describing the heavens. Most importantly of all, Galileo discovered the power of mathematics in describing the natural world. Nature, he wrote,

> is written in this grand book, the universe, which stands continually open to our gaze. But the book cannot be understood unless one first learns to comprehend the language and read the letters in which it is composed. It is written in the language of mathematics, and its charac-

ters are triangles, circles, and other geometric figures without which it is humanly impossible to understand a single word of it; without these, one wanders about in a dark labyrinth.

The importance of this idea cannot be stressed enough; indeed, it will emerge again and again over the course of our story. There is order in nature and that order can be described through mathematics, by means of simple, concise—even elegant—equations. Experiment and mathematical analysis, taken together, would serve as the backbone of science. We credit Galileo with the birth of this idea, a notion that would find its greatest champion in the work of an Englishman—a lad from Lincolnshire born just a few months after Galileo's death.

Isaac Newton: The Lone Genius

Isaac Newton (1642–1727) was born in the rural manor house of Woolsthorpe on Christmas Day, 1642. Described by a family friend as a "sober, silent, thinking lad," Newton lived at Woolsthorpe until the age of 12, when he began his studies at the grammar school in Grantham. The old stone schoolhouse, built in 1528, still stands; a plaque on the outer wall commemorates the school's most illustrious pupil. Inside, a stone window ledge is covered with seventeenth-century graffiti—including the engraved signature of a certain "I. Newton." In the town square stands a massive bronze statue of Newton—just across the street from the Isaac Newton Shopping Centre.

We know little about Newton's school days, but a fight he had with the school bully appears to have had a profound impact on him. The bully had kicked him in the stomach, prompting Newton to challenge the boy after class. John Conduitt, who would later marry Newton's niece, gives this account: "Tho Sir Isaac was not so lusty as his antagonist, he had so much more spirit and resolution that he beat him until he declared he would fight no more, upon which the schoolmaster's son bade him use him as a coward, and rub his nose against the wall, and accordingly Sir Isaac pulled him along by the ears and thrust his face against the side of the church." The fight apparently kick-started Newton's academic performance: before the skirmish, he was near the bottom of his class; afterward, he rose to be first in the school.

Newton's mother hoped Isaac would eventually return to Woolsthorpe to look after the family estate. But Newton was clearly unsuited to the task, his mind forever wandering from matters at hand. He preferred whiling away the hours building models of clocks and windmills rather than helping tend the crops and animals. Finally, his mother accepted that he wasn't cut out for a farmer's life, and allowed him to enroll at Cambridge. The servants were apparently pleased; they're said to have "rejoiced at his departure, declaring he was fit for nothing but the 'Versity."

Newton went off to Cambridge in the summer of 1661. The great scientific discoveries of the previous century were not yet taught—the university was still a strictly Aristotelian world—but Newton had heard about the revolutionary new ideas that were changing the face of natural philosophy. He had read about the astronomy of Copernicus and Kepler, the mechanics of Galileo, and the philosophy of Descartes, whose view of nature as an intricate but predictable mechanical system deeply influenced him.

For Newton, ideas seemed to hold more appeal than living people. He had few close friends in Trinity college, and rarely dined in the Great Hall—and when he did, he often "would go very carelessly, with shoes down at heels, stockings untied ...and his head scarcely combed." Later, as a faculty member, he was the very model of the absent-minded professor; on one occasion he's said to have lectured to an empty room.

Newton spent long hours alone in his study, scribbling his ideas by candlelight and conducting experiments, often well into the morning, in a small laboratory adjacent to his rooms. His

The English physicist and mathematician Isaac Newton. His *Principia* marked the climax of the Scientific Revolution.

greatest insight, however, came not at Cambridge but back home at Woolsthorpe. For 18 months, beginning in 1665, an outbreak of the plague forced the university to shut down, and Newton—now in his early 20s— returned to the family home. "In those days I was in the prime of my age for invention," he would later recall, "and minded mathematics and philosophy more than at any time since." During those tranquil days in Woolsthorpe, Newton mastered the laws of mechanics, elaborating on Galileo's work and establishing what are now called Newton's three laws of motion—laws that still form the foundation of classical physics. He also laid the groundwork for his invention of calculus and began his investigation of light and color.

It's the story of the apple, however, that's taken on legendary proportions in the annals of science. Lazing in the garden of his boyhood home, Newton—no doubt, lost in his thoughts—saw an apple fall to the ground. As he contemplated the falling apple, he also thought about the moon moving in its orbit around the earth—and came to a profound conclusion. He knew it was gravity that tugged on the apple; perhaps gravity held sway even at vastly greater distances. As he recalled later, he "began to think of gravity as extending to the orb of the moon."

Newton set to work finding the equations that governed such motion. Finally he "deduced that the forces which keep the plan-

Newton deduced that the same force—gravity—governs a falling apple and holds the moon in its orbit.

ets in their orbs must [be] reciprocally as the squares of their distances from the centers about which they revolve." Or, in modern language, the strength of gravity follows an *inverse-square law*: double the distance between two bodies and the gravitational force between them falls to one-quarter of its value. Triple the distance, and the force is reduced to one-ninth of its strength. Using the inverse-square law, Newton showed that planetary orbits *had to be* elliptical; one follows mathematically from the other. He could use his laws to derive those of Kepler. And it didn't matter whether the bodies involved were planets, moons, or apples: gravity worked in exactly the same way in every case, and the same mathematical description could be applied to all. The law of *universal gravitation* was born.

Of course, the story of the apple, which Newton only told very late in his life, should be taken with a grain of salt. Like the story of Galileo and the Tower of Pisa, it was told by an aging scientist to a hero-worshipping disciple many years after the fact. And, like the Pisa story, the truth of the anecdote hardly matters today. But the visitor to Woolsthorpe still encounters an aging, twisted apple tree on the gently sloping lawn in front of the manor house. Although much too young to be *the* tree, its weathered trunk and gnarled branches still hold an irresistible appeal.

Newton's *Principia*: Laying the Foundation

Join me in singing the praises of Newton,

Who opens the treasure chest of hidden truth ...

No closer to the gods can any mortal rise.

EDMOND HALLEY, IN THE PREFACE TO NEWTON'S *PRINCIPIA*

Though Newton strove for seclusion, word of his achievements inevitably spread. In 1669, at the age of 27, he was appointed to a prestigious academic position in Cambridge, becoming the Lucasian Professor of Mathematics. (The post is currently held by Stephen Hawking, famed author of *A Brief History of Time*.) Meanwhile, Newton's work in optics led to an invention of great practical value—a telescope that used mirrors rather than lenses to form images. (Telescopes with lenses, like those of Galileo, are called *refractors*; they are expensive to produce and sometimes fail to deliver true colors. A large *reflector*, by comparison, is relatively easy to make and delivers sharp hues.) When scientists at the newly-established Royal Society of London got word of Newton's invention, they invited him to join their ranks; he later led the prestigious society as its president.

Still, Newton was hesitant to publish many of his results. The work that eventually became his magnum opus collected dust in his study for more than 20 years, until astronomer Edmond Halley (later of Halley's comet fame) urged him to publish. The massive volume, which set out his theories of motion and gravity, was finally printed in 1687 under the weighty title *Philosophae Naturalis Principia Mathematica* (*Mathematical Principles of Natural Philosophy*). Instantly hailed as the most important treatise on physics ever written, it immediately established Newton as Europe's foremost man of science.

Newton spent more than 35 years at Cambridge. Finally, in his fifties, he felt the university had little left to offer him. He moved to the capital and accepted the post of Warden, and later Master, of the Royal Mint. His London years also found him mired in conflict with his fellow scientists. The most virulent of these was with German mathematician Gottfried Leibniz (1646–1716) over the invention of calculus—a quarrel which dragged on for more than a decade and ended only when Newton learned of his adversary's death. (Today, most historians believe the two men developed calculus independently.)

Knighted by Queen Anne in 1705, Newton was, in his final years, Britain's supreme scientist. When he died in London at the age of 85, he was given a state funeral and the ultimate British honor—burial in Westminster Abbey. Above the tomb is an ornate marble sculpture of the reclining Newton, accompanied by a globe, cherubs, and a female figure representing Astronomy, the Queen of the Sciences. The Latin inscription reads, "Let Mortals rejoice That there has existed such and so great an Ornament to the Human Race."

Scientist or Sorcerer?

Newton's achievements in science—his work in optics, his invention of calculus, his theories of motion and gravity—are unparalleled. In the wake of his discoveries, people began to see nature as a well-tuned machine, a clockwork governed by strict mathematical laws. That metaphor, anchored in his mammoth *Principia*, dominated scientific thought for more than two hundred years. But Newton lived at a time when the echo of the medieval world, though fading, could still be heard. He wrote treatises on ancient history, mythology, Biblical chronology, and a host of other arcane subjects; among his obsessions was an attempt to determine the date of Armageddon by a careful study of the Book of Daniel. Modern scholars talk about his work in chemistry, but it was really the ancient art of *alchemy* that fascinated Newton. Biographer Michael White points out that Newton had 138 volumes on alchemy in his library at the time of his death, compared with 31 on what we would call chemistry.

What drove Newton—remembered today as a man of logic and rationality—into alchemy and the occult? Like today's physicists, he sought a unified, all-encompassing description of nature; unlike modern scientists, however, he was afraid to exclude the ideas of ancient scholars and philosophers. According to White, Newton "was interested in a synthesis of all knowledge and was a devout seeker of some form of unified theory of the principles of the universe." Such knowledge, Newton believed, was at one time known to philosophers, but had since been lost; his great quest was to rediscover this ancient wisdom. "Newton's *raison d'être* was to recover this 'frame of knowledge,'" writes White. "For this reason, on an intellectual level, he considered no avenue of research beyond his probings, no stone unimportant enough to be left unturned, no theory beyond the pale." Newton would probably have disliked being labeled as either a chemist or alchemist; he was simply a natural philosopher searching for the most elementary laws of nature, laws that would define a unified theory. He was looking for a Theory of Everything.

Newton's alchemical investigations proved futile, yet within physics his quest for unification was a spectacular success. The elliptical orbits of the planets, the rise and fall of the ocean's tides, the paths of projectiles and falling bodies, even the shape of the earth itself—all could be explained through Newton's succinct mathematical laws. Where many theories were once employed, now only two ideas were needed: Newton's laws of motion and his law of universal gravitation. He had cemented the unification of terrestrial and celestial physics begun by Galileo. Newton had constructed a mathematical framework for the physical sciences that held within it all the great advances of the previous millennium and set the stage for the progress to come.

And, like the Greeks and the scientists of the Renaissance, Newton sought simplicity. Echoing the words of William of Ockham, he wrote in his *Principia*: "Nature does nothing in vain, and more causes are in vain when fewer suffice." He believed that "Nature is simple and does not indulge in the luxury of superfluous causes."

Newton may indeed have mixed reason and magic—but he was also the first giant of the new scientific age, the culminating figure of the Scientific Revolution. Following Newton, science was finally weaned from philosophy and transformed into a discipline of inquiry in its own right. Those seeking the truth no longer needed to rely on the insights of a few learned men or the authority of a handful of dusty textbooks—instead, insight would come from direct observation of nature, carried out through precise measurement and mathematical analysis. That strategy, inaugurated by Galileo and firmly established by Newton, is now called the *scientific method*.

With the work of Galileo and Newton, science had reached maturity. Galileo, the first to realize the enormous power of mathematics in describing nature, is recognized as the father of modern physics—and so we honor him with a T-shirt: it would show the equations that govern falling bodies and accelerated motion. In recognition of his groundbreaking work in astronomy, perhaps a drawing of Jupiter and its moons could go on the back. We could even use a photograph from the Galileo space probe. And Newton brought it all together in one concise, simple package. Any first-year physics student could design his T-shirt: it would display Newton's three laws of motion and his law of gravity—laws that apply equally in the heavens and on the ground. Galileo and Newton had completed the first great breakthrough in the quest for a unified theory of physics. They had linked heaven and earth.

Flashes of Insight

Electricity, Magnetism, and Light

It is presumed that there exists a great unity in nature, in respect of the adequacy of a single cause to account for many different kinds of consequences.

IMMANUEL KANT

For two centuries following the publication of Newton's *Principia*, the mechanical world-view held sway. Scientists eagerly applied Newton's laws to problems in astronomy, physics, and engineering with spectacular success. Perhaps the best example is a prediction made by the English astronomer Edmond Halley. Observing a bright comet in 1682, Halley was able to use Newton's laws to work out its orbit, and predicted that it would return 76 years later. And sure enough, the space-faring chunk of rock and ice reappeared in the sky in 1758, right on cue. (Halley died in 1742, but is immortalized in the comet that now bears his name.)

Yet the mechanical picture, described by Newton's law of gravity and his laws of motion, didn't seem to cover everything. In particular, the nature of electrical and magnetic forces was still a mystery. What little was known, in fact, had been known to the ancient Greeks two millennia earlier. For example, Thales of Miletus (one of our heroes from the first chapter) had noted that certain black rocks had the power to attract metals such as iron; the Greeks called these stones *magnets*, after the region of Magnesia, in Asia Minor, where they were commonly found. They also knew that rubbing certain materials, like amber, caused them to attract lightweight objects like cork, paper, or bits of hay.

In the Middle Ages, someone discovered that if a lightweight magnet were suspended so that it was free to rotate, it aligned itself in a north-south direction—and the mariner's compass was born. (The discovery dates from around the eleventh century, and may have originated in China.) And yet, in terms of the underlying theory, progress was painstakingly slow. In 1600, the English scientist William Gilbert (1544–1603), physician to Queen Elizabeth, summed up what was then known about electricity and magnetism in a grand treatise, *de Magnete*. Gilbert claimed—correctly—that the earth itself was a natural magnet; he also coined the word *electricity* from the Greek word for amber, *elektron*.

However, Gilbert's attempts to link electricity and magnetism to other forces of nature failed. Like Kepler, Gilbert sought a link between magnetism and the motion of the planets around the sun. It was, on the surface, plausible enough; he had already shown that the earth had a magnetic field, and we know today that the sun and many of the other planets do as well. With the work of Newton, however, it became clear that gravity, not magnetism, governed planetary motion.

After Gilbert, another two centuries slipped by with little progress in the realm of electricity and magnetism. The slow pace of discovery is perhaps not surprising. First, in the wake of Newton's great discoveries, both electricity and magnetism were seen as intriguing but not particularly important phenomena; secondly, until the dawn of the nineteenth century, the tools needed to study them in detail simply did not exist.

Electricity and Magnetism: A Subtle Connection

Some kind of link between electricity and magnetism had long been suspected, but the evidence was mostly anecdotal. In 1731, for example, lightning struck the kitchen of an English tradesman; when the dust had settled, he found that some of the knives and spoons had the power to pick up nails and other small bits of iron: they had become *magnetized*. In 1752, the American inventor and statesman Benjamin Franklin (1706–90) flew a kite during a lightning storm—and demonstrated the link between lightning and electricity. (Had Franklin hosted a spot on the Discovery Channel, he would no doubt have cautioned, "Don't try this at home.")

But electricity is a slippery subject: you can't study what you can't store, and electric charges have a way of dissipating before they can be measured and analyzed. For many years, the only way to store electric charge was with a "Leyden jar"—a sealed glass container lined with metal. The first scientists to examine charges quantitatively began the study of *electrostatics*—the forces between stationary charges. They quickly realized there were two kinds of charge; we call them simply "positive" and "negative." Opposite charges were seen to attract one another, while similar charges repelled. Before long, an inverse-square law (analogous to that of gravity) was found to govern the strength of the force between charges. The rule is usually called Coulomb's Law after the French scientist Charles-Augustin de Coulomb (1736–1806); however, Joseph Priestley (1733–1804), a British clergyman and chemist known primarily for his work with gases, independently discovered the inverse-square law at about the same time. Another Englishman, Henry Cavendish (1731–1810), also made important contributions to electrostatics, though he's better known for isolating the element hydrogen and measuring the strength of gravity with great precision.

The next breakthrough came, as sometimes happens in science, by sheer accident. In 1786, the Italian physiologist Luigi Galvani (1737–98) touched the leg of a dissected frog with an electrical charge—and observed a violent contraction. He thought the effect originated in the animal's organic tissue, but it was actually the salt within the tissue, in concert with Galvani's metal electrodes, that was responsible. His discovery led to the invention of the electro-chemical battery. Another Italian, the physicist Alessandro Volta (1745–1827), took the next step. In 1800, he produced the "voltaic pile"—a stack of alternating layers of silver, zinc, and cardboard which, when placed in an electrical circuit, produced a continuous stream of electricity. The quantitative study of *electric current* had begun.

The Emperor Napoleon, sensing the promise in this string of discoveries, called for more research, suggesting that prizes be awarded for advances in understanding electricity. "Galvanism, in my opinion, will lead to great discoveries," he declared. And indeed it did, although 20 more years passed before another chance discovery illuminated the elusive link between electricity and magnetism.

Hans Christian Oersted: The Current and the Compass

Hans Christian Oersted (1777–1851), the son of an apothecary, was born on the Danish island of Langeland. He quickly took in as much knowledge as his small hometown could offer. His father, though poor, recognized the boy's voracious appetite for reading and learning, and managed to enrol him at the University of Copenhagen. At first Oersted studied pharmacy, until a fascination for physics took over; he was later appointed full professor.

Oersted heard about the recent discoveries involving electric current—including the groundbreaking work of Galvani and Volta—and soon began investigating the electrical properties of acids and metals. In 1820, Oersted was preparing a lecture demonstration on the effect—if any—of an electric current on a magnetic compass needle. He only had time to investigate a few of the many possible arrangements of the apparatus, so he was still unsure of the results as he entered the classroom full of students. When the current-wire was placed beside the compass at the same height as the compass, there was no effect. But when it was placed above or below, the compass needle suddenly moved. Intriguingly, the needle did not line up parallel to the wire, but at right angles to it—a result no one at the time could have foreseen. It was, historians have mused, probably the greatest scientific discovery ever made during a lecture demonstration—and one that would have consequences far beyond anything Oersted could have imag-

The Danish physicist Hans Christian Oersted. In what may have been a lucky accident, he discovered that a current flowing in a wire generates a magnetic field.

Science Photo Library/Publiphoto

ined. The long-suspected connection between electricity and magnetism was now demonstrated, and a technological revolution would soon follow.

Oersted published his results in a four-page tract in July, 1820—one of the last important scientific papers to be penned in Latin. Similar experiments were soon performed, with the same result, in laboratories across the continent. News of Oersted's discovery reached Paris in a matter of weeks. André Marie Ampère (1775–1836), a French mathematician and scientist, was in the audience when Oersted's discovery was announced before l'Académie des Sciences. Ampère was inspired to begin his own investigations— the first detailed studies in *electrodynamics*. He quickly discovered that current-carrying wires attracted or repelled one another, depending on the direction of the current. The strength of the force again obeyed an inverse-square law, just as it had for gravity and electrostatics. Just one week after hearing the news of Oersted's discovery, Ampère was ready to present his own paper to the academy. The French physicist François Arago was stunned by the speed of Ampère's work: "The vast field of physical science perhaps never presented so brilliant a discovery, conceived, verified and completed with such rapidity."

Back in Copenhagen, Oersted was hailed as the greatest Danish scientist since Tycho Brahe. Though remembered for his great discovery, Oersted is also an icon in the world of teaching: today, the highest honor awarded to American physics teachers is the Hans Christian Oersted Medal. Toward the end of his life, he wrote a series of essays on the harmony he saw between beauty and science; both, in his eyes, were God's work. "Spirit and nature are one, viewed under two different aspects," he wrote. Like the ancient Greeks, like Kepler and Newton, he sought unity in nature. His discovery—finding the link between electricity and magnetism—was another giant step toward a Theory of Everything. The next breakthrough would come from an Englishman who grew up in poverty but went on to become one of Britain's greatest scientists.

Michael Faraday: The Master Experimenter

Michael Faraday (1791–1867) is one of the great romantic figures in the history of science; the story of his life is a true "rags to researcher" tale. Faraday was the third child of a poor blacksmith. His family lived in the downtrodden London neighborhood of Newington Butts, not far from what is now the Elephant and Castle underground station. With little education, Faraday was sent to work in a bookbinder's shop at the age of 14. Though occupied with manual labor, after hours Faraday was allowed to read the books that were bound in the shop; those on science held his attention most of all.

In 1812, a customer took Faraday to hear a lecture by Sir Humphrey Davy at the Royal Institution, a society founded in 1799 to promote public dissemination of applied research. Davy, struck by the boy's enthusiasm, later made him an assistant at the Institution. For Faraday, it was a dream job; he soon began his own experiments in electromagnetism. By December, 1821—less than 18 months after the publication of Oersted's results—he had invented a primitive version of the electric motor. In what seems like the start of a "to do" list, he then made a note in his diary: "Convert magnetism into electricity!"

British scientist Michael Faraday. His discovery that magnetism could be used to generate an electrical current sparked a technological revolution.

Science Photo Library/Publiphoto

A decade later, Faraday succeeded. In August, 1831, he discovered how to use magnetism to induce an electrical current in a wire, essentially the reverse of what Oersted had done. A static magnetic field, Faraday discovered, would not do the trick, but a *changing* field indeed made a current flow in a nearby circuit. (The actual strength of the field could be made to vary, or the source of the magnetism, whether a traditional magnet or a current-loop, could be set in motion.) Faraday had discovered *electromagnetic induction.* It would lead to the first electrical transformers and dynamos—and to an explosion in the development of electrical technology.

Faraday never patented anything he invented. A modest, deeply religious man, he was interested only in the underlying principles behind his discoveries. Yet his skill as a public lecturer and science communicator assured his fame. (During the 1990s, Faraday's portrait graced the British £20 banknote; he was later replaced by the composer Edward Elgar.) Faraday was adamant that science be conducted for its own sake and not be driven by the immediate demands of industry—a view that many of today's scientists would sympathize with. With almost no training in mathematics, however, he could only go so far; the Scottish physicist James Clerk Maxwell would complete the electromagnetism revolution, as we'll see shortly.

Faraday, however, was probably the first to realize the importance of a new concept in physics. For nearly two centuries, physicists had been thinking solely in terms of *forces* acting between

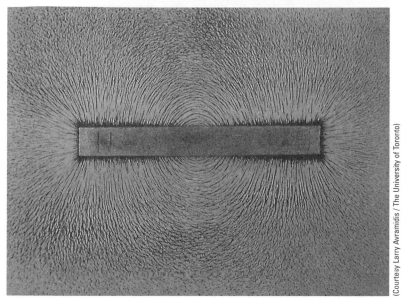

Iron filings sprinkled above a standard bar magnet reveal the invisible magnetic field. The field lines curve from the magnet's "north" pole to its "south" pole.

(Courtesy Larry Avramidis / The University of Toronto)

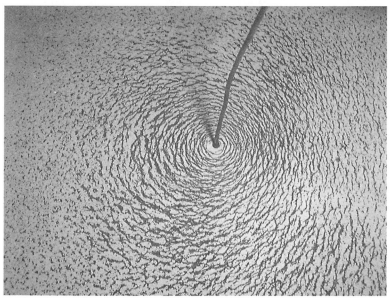

A current flowing through a wire also generates a magnetic field—this time, in a circular pattern.

particles; now they began to think in terms of the *field*. An electric charge, for example, is surrounded by an electric field; the closer one is to the source, the stronger the field. The same idea applies to magnetic and gravitational fields. The simplest demonstration is the one we all saw in school, when the teacher sprinkled iron filings on a sheet of paper held above a bar magnet. The resulting pattern reveals the unseen but ever-present magnetic field. Fields would soon play a vital role in almost every branch of physics, allowing for simplified calculations and the development of more abstract and more powerful theoretical models. Yet Faraday never saw his work as revolutionary. "Do not suppose that I was a very deep thinker," he wrote, "but facts were important to me."

A host of other scientists were more than willing to investigate the technological applications of Faraday's work, applications he himself shunned. The results came swiftly. The American physicist Joseph Henry patented the electric motor in 1835. Three years later, an electric railway was running between Glasgow and Edinburgh. In 1843, Samuel Morse—of Morse code fame—built a telegraph line from Washington to Baltimore. With the help of undersea cables, the electric telegraph soon connected every corner of the globe. Electric trams, trains, and elevators—along with reliable electric lighting—transformed cities and changed the way people live.

While the Copernican revolution had a profound impact on the way humanity viewed itself, it had little effect on people's day-to-day lives. By comparison, the discovery of electromagnetism had immediate practical consequences. When Faraday was born, messages traveled no faster than horses or ships could carry them; by the time of his death, information could be sent across continents and oceans as fast as it could be typed or read.

Electromagnetism, as a science, did not exist before 1820; by the end of the century it had changed the face of the world.

James Clerk Maxwell: The Great Synthesis

James Clerk Maxwell (1831–79) was born in Edinburgh to an established upper-middle-class family, just a few months before Faraday's discovery of electromagnetic induction. Maxwell was a precocious, inquisitive child, who loved exploring the courses of rivers and streams—and of the bell-wires that ran through the house (used to summon the servants). Playing with a highly polished silver tray, the two-year-old Maxwell thought that he had invented the mirror—a precursor, perhaps, to his groundbreaking work on the nature of light. Above all, he constantly asked questions—about nature, about the sun and stars, about beetles and frogs, about rocks and metal. An aunt once remarked, "It was humiliating to be asked so many questions one couldn't answer by a child like that."

Maxwell wrote his first scientific paper, on the geometry of certain ellipse-like curves, at the age of 14. Suddenly, commentators were speaking of this astute Scottish teenager in the same breath as Descartes and Newton. Two years later he entered the University of Edinburgh. Fascinated by the revolution in electromagnetism, he studied the work of Oersted, Faraday, and the other pioneers in this exploding field of study.

In 1850, Maxwell went off to Cambridge, first as a student and eventually as a professor. He later held teaching positions at Kings College in London and the University of Aberdeen in his native Scotland. (The latter institution, he said, was the most uptight. "No jokes of any kind are understood here," he complained. "If I feel one coming, I bite my tongue.") But it was in Cambridge that he began his great work on electromagnetism.

The British physicist James Clerk Maxwell. His equations of *electromagnetism* govern all known electrical and magnetic phenomena.

While Faraday knew virtually no mathematics, Maxwell was a mathematical prodigy, a man more than comfortable in the world of differential equations and abstract geometrical ideas. Maxwell exploited a newly developed branch of mathematics known as vector analysis (or vector calculus) in his physics investigations. When necessary, he extended those mathematical tools and developed new ones.

Just as Kepler had used his mathematical skills to find the underlying patterns in Tycho Brahe's astronomical data, Maxwell used his numerical knowhow to find the patterns lurking in Faraday's work on electricity and magnetism. In December, 1855, he presented the beginnings of a theory to a meeting of the Cambridge Philosophical Society, in a paper titled "On Faraday's Lines of Force." Faraday later wrote to Maxwell: "I was at first almost frightened when I saw such mathematical force made to bear upon the subject, and then wondered to see that the subject stood it so well."

Over the next 20 years, Maxwell developed and finally perfected his theory. In a series of lectures between 1861 and 1865, he presented the four equations that now bear his name—the famous "Maxwell's equations," which describe the relationship between electric and magnetic fields. His summary, the *Treatise on Electricity and Magnetism* of 1873, unified all of the known facts about electricity and magnetism into one concise theoretical framework. This was the second great revolution in physics. Just as Galileo and Newton had linked celestial and terrestrial mechanics,

Maxwell had shown that electricity and magnetism—and even light and optics, as we'll see shortly—are all intimately related. As biographer Ivan Tolstoy put it, Maxwell's theory is, "first of all, a synthesis—one of the greatest in the history of science ...a whole universe of electromagnetic phenomena, miraculously contained in a few lines of mathematics."

Maxwell's work on electromagnetism is so important that it overshadows his many other achievements in science. He made important contributions to thermodynamics, the molecular theory of gases, color vision, and even color photography. He also conducted one of the first theoretical studies of the mechanics of Saturn's rings. Maxwell even tried his hand at verse; many of his poems reveal the scientist as a sensitive, witty observer of life.

Though honored among scientists as the greatest physicist after Newton and Einstein, Maxwell never became a household name. Throughout his life, he remained a modest, extremely private man. Like Faraday, he shunned publicity and public honors. And, like Faraday, he was very religious. (We'll hear more on the relationship between science and religion in the final chapter.) Sadly, Maxwell died at 48 from abdominal cancer, a disease that had killed his mother at the same age.

Let There Be Light

As we've seen, Maxwell's work brought together electricity and magnetism under one unified description. In doing so, it also illuminated—forgive the pun—a subject that had long puzzled physicists: the problem of light. Like space and time, light is one of those paradoxical entities that at first seems to defy any kind of scrutiny. It appears as a fundamental part of our universe, seemingly elementary—and yet for centuries it eluded direct study. What, scientists wondered, was light actually made of?

Disturbances on the surface of a pond—in this case, from two tossed stones—cause waves to spread out in circular patterns across the water. Light works in an analogous way, radiating from point sources in spherical waves.

Isaac Newton, inspired by the success of mechanical models in so many other areas of physics, looked for a mechanical description of light. Rays of light must be made of particles, he said, which stream forth from the sun, flames, and other luminous objects. That explained some phenomena associated with light but not others. The Dutch physicist Christiaan Huygens (1629–93), on the other hand, said light was a wave, similar to the more familiar water waves and sound waves. The wave nature of light received further support from the experiments of the British physician and scientist Thomas Young (1773–1829) and the French physicist Augustin Fresnel (1788–1827), who investigated a wide range of optical phenomena and showed, among other things, that light waves could interfere with one another, just like the water waves seen in the photograph above.

Enter James Clerk Maxwell, whose theory of electromagnetism seemed to *require* the existence of a certain kind of wave—a self-propagating, periodic disturbance in an electromagnetic field (what became known as *electromagnetic waves*). Further, one could make certain electrical and magnetic measurements that would yield the precise value for the speed at which these waves traveled. And it turned out—you guessed it—that they traveled with a speed that exactly matched the known speed of light (within experimental errors at the time). "We can scarcely avoid the inference," Maxwell wrote, "that *light consists in the transverse undulations of the same medium which is the cause of electric and magnetic phenomena*" (Maxwell's italics).

Proving that visible light was the result of oscillating electromagnetic fields was, at that time, not technically feasible—but there was another way to see if the idea was correct. If Maxwell was right, any rapidly oscillating electric field—even one with a very low frequency—should also produce waves—not waves of visible light, but less intense vibrations of a much longer wavelength. In 1887, eight years after Maxwell's death, German physicist Heinrich Hertz (1857–94) finally produced and measured these longer *radio waves*. This was the ultimate triumph for Maxwell's theory. Hertz also showed that radio waves could be reflected and refracted just like light waves. Light, in other words, was just one type of *electromagnetic radiation*. The impact of Hertz's findings, like that of Faraday's research, was swift. Less than a decade later, the Italian physicist and inventor Guglielmo Marconi (1874–1937) built the first wireless telegraph machine—the precursor of the radio.

The end of the nineteenth century was one of the high-water marks in the history of science—a time of enthusiasm, optimism, and an unshakable sense of progress. Thanks to the work of Oersted, Faraday, and Maxwell, a remarkably coherent picture of the physical world was emerging. However, this view had certain shortcomings—some of them quite serious—which would soon lead physics in new directions. But that should not stop us from celebrating this incredibly rich period of scientific questioning, experiment, and discovery.

The discovery of the laws of electromagnetism was a giant leap on the path toward a simple, concise description of nature. Maxwell's work in particular stands out; his achievement ranks alongside that of Charles Darwin as the most significant scientific advance of his century. Maxwell's synthesis, embodied in his famous equations, sums up everything we know about electric and magnetic fields. Indeed, by explaining light as an electromagnetic wave, it also embraces every kind of optical phenomenon, from the reflection and refraction of light to interference effects—as well as their counterparts in radiation of other wavelengths, from X-rays to radio waves. These waves had always existed, of course, but they lay hidden until a series of inspired thinkers, from Oersted to Maxwell, brought them to light. The choice of T-shirt is a simple one: just about any university with a physics or engineering department will sell you a T-shirt displaying Maxwell's equations.

But it was the technological innovations spawned by Maxwell's discovery, rather than the theory itself, that changed the world so dramatically and so quickly. Electric appliances, lights, radio, television, telephones, computer networks, satellite communication—these innovations now define the very fabric of our lives; we can hardly imagine an age in which they did not exist. Yet it all began with Oersted's lecture-hall discovery that an electric current makes a compass needle swing.

Something else changed during these decades: science began to retreat from the everyday world of the senses. Gravity and the laws of mechanics, and their myriad effects, are with us every moment of our lives, usually in very obvious ways. Step off a ladder and you'll soon be reminded of gravity. Shoot a hockey puck down the rink or sink the eight ball in a game of billiards, and you're demonstrating Newton's laws of motion. Probing electricity and magnetism, however, requires more sophisticated tools. By the close of the nineteenth century, science no longer dealt only with objects and motions that we could see or touch directly. It was now a more abstract discipline, one that demanded specialized knowledge and techniques and one that required, in most cases, high-level mathematics. Once open to the interested amateur, science had become the exclusive domain of scientists. Ordinary members of the public could still taste the excitement of the latest discoveries, and could take advantage of the proliferating technology that they led to—but the details had to be left to the experts.

No one embodies this transformation more than the man who would revolutionize physics in the first two decades of the twentieth century—a man who embodies the spirit of scientific discovery and mathematical insight, but whose work, and even his name, have become nearly synonymous with "unfathomable." His name, of course, was Albert Einstein. As we'll see in the next chapter, however, his theory of relativity was in fact a natural continuation of the work of Oersted, Faraday, and Maxwell. He would, of course, take physics in directions that those men could not have envisioned; yet his goal was the same as theirs: to explain the largest number of facts with the fewest assumptions. He would continue the quest for the Theory of Everything.

Relativity, Space, and Time

Einstein's Revolution

Absolute, true, and mathematical time, in and of itself and of its own nature, without reference to anything external, flows uniformly....Absolute space, of its own nature, without reference to anything external, always remains homogeneous and immovable.

ISAAC NEWTON, 1687

Henceforth space on its own and time on its own will decline into mere shadows, and only a kind of union between the two will preserve its independence ...

HERMANN MINKOWSKI, 1908

"If I had to sum up the twentieth century," musician Yehudi Menuhin said shortly before his death in 1999, "I would say that it raised the greatest hopes ever conceived by humanity, and destroyed all illusions and ideals." That turbulent century saw change and upheaval on a magnitude never before imagined. It was a century of contradiction, giving us trench warfare, genocide, environmental destruction, and the peril of overpopulation—but also bringing freedom and prosperity to millions and leaving us with a bounty of scientific and technological breakthroughs and a new and deeper understanding of our universe.

When, exactly, did this new world arrive? For historian Eric Hobsbawm, the starting point was the dawn of the first global war; his choice is reflected in the title of his book, *The Age of Extremes: The Short Twentieth Century 1914-1991*. Other scholars, of course, point to a variety of alternative beginnings. Perhaps the modern world began when the dissonant chords of Igor Stravinsky's *The Rite of Spring* sounded in a Paris theater on May 29, 1913, prompting a near-riot from the audience. Or maybe it began in the summer of 1907, when Pablo Picasso unveiled his shocking vision of five prostitutes, *Les Demoiselles d'Avignon*, with its distorted figures and harsh geometrical lines. I would vote, however, for an event that took place two years earlier. On June 30, 1905, a German physics journal published a paper by a young scientist working at the Swiss patent office in Bern. It was Albert Einstein's first paper on relativity—and it changed the world forever.

To understand why Einstein's work had such a profound impact on our view of the universe, we have to take a closer look at where physics stood at the end of the nineteenth century. Newton's laws, though more than 200 years old, were going strong. Some of the implications of those laws, however, were troubling. Consider his law of universal gravitation: the law states that every object in the universe exerts a gravitational tug on every other object, no matter how far apart they are. But how, exactly, is this force transmitted? When you pull a toboggan along the snow, you're always in contact with it, or at least with a rope that's tied to it. But gravity, somehow, leaps across empty space. It allows the sun, for example, to hold the earth in orbit even though there is no physical connection between the two. The best guess that Newton's followers could come up with was that the force of gravity was transmitted via a peculiar, invisible substance that pervades all of space. They called it the "ether."

This mysterious ether was still on physicists' minds when Maxwell unveiled his laws of electricity and magnetism in the nineteenth century. In the wake of Maxwell's work, light was seen as an electromagnetic wave—a series of ripples in an electromagnetic field. But light waves were obviously very different from water waves or sound waves. If you throw a stone in a pond, waves spread out across the water, radiating away from the point of impact. If you clap your hands, sound waves travel outward through the air in the same fashion. Neither wave would exist without the medium through which it travels: water waves need water; sound waves demand air. Now, light was found to be a wave—but through what medium did it travel? Physicists again needed to invoke the ether.

This ether, however, was a very strange substance. It must fill the entire universe, even space that is seemingly empty; after all, light is clearly able to reach us from the sun and distant stars across the near-vacuum of deep space. The ether must be invisible and massless, offering no resistance as the earth plows through it in its orbit around the sun. (Otherwise, there would be friction between the earth and the ether; this would slow our planet and make the earth gradually spiral toward the sun.) Light, however, made other demands on the ether: in order to transmit electromagnetic waves, ether must have the properties of what physicists call an "elastic solid." In other words, it had to be extremely rigid—even more rigid than steel. Nobody knew how such an exotic substance could possibly work. The ether also seemed to be necessary in order to explain something rather puzzling about the speed of light. To understand this last point, we'll have to examine our ideas about motion and speed.

Relativity before Einstein

Long before Einstein, physicists understood that motion is "relative"—that is, any measurement of speed makes sense only in reference to some other object. Sometimes this ambiguity can cause confusion, as anyone who's traveled by train has experienced: while your train is in the station, you look out the window and see another train on the next track. Suddenly, it begins to move. But wait a moment: is it really moving, or has your own train begun to move? This simple idea of relative motion was understood even in ancient times. "We sail forth from the harbor, and lands and cities draw backwards," the Roman poet Virgil wrote in the first century B.C.

By Galileo's time, it was well known that measurements of speed had to take into account both the speed of the object and the speed of the observer. Here's a simple example. Suppose you're on board a train that's moving at 60 kilometers per hour, and you throw a baseball forward with a speed of 80 kilometers per hour. As seen by an observer on the ground, the baseball has a speed of 140 kilometers per hour. Easy, right? You just add the two speeds together. Isaac Newton used this principle as a cornerstone of his laws of motion. Space and time were absolute and unchanging, he said, while different objects moved at different speeds against this fixed backdrop. The idea of relative motion, along with absolute space and time—anchored by the all-pervasive ether—appeared beyond refute. It accorded with our "common sense" picture of the world so well, in fact, that for more than 200 years it was never questioned.

Then we come to Maxwell. His equations for electromagnetism predicted that light was endowed with a specific, constant speed—about 300,000 kilometers per second—but, relative to what? Again, only one answer seemed to make sense: light must travel at a constant speed relative to the ether. If that were the case, however, then the earth's motion through the ether should affect our measurements of the speed of light.

One of the first efforts to detect such a variation was carried out by two American scientists, Albert Michelson (1852–1931) and Edward Morley (1838–1923). Their idea was simple: if the earth was moving through the ether in some particular direction, then light should travel at one speed parallel to that direction, and at another speed in the perpendicular direction. Because the earth's speed is so small compared with the speed of light, the difference would be very slight—but the Michelson-Morley apparatus was sensitive enough to detect it. Incredibly, they found no such difference. The earth's motion had no effect at all on the speed of light. Just to be certain, they repeated the experiment at different times of the year, when the earth's orbital motion carried it in different directions—but it made no difference. There was simply no sign of the earth's motion through the ether.

For more than a dozen years, that's where the matter stood. More experiments were carried out, but no one found any sign of the ether—and yet both Newton's gravitation and Maxwell's electromagnetism seemed to demand it. It would take a new way of thinking to resolve the impasse, and, in a moment, we'll see what that solution was. But first we will meet the man responsible for this new perspective—a man who was not afraid to question the established view of physical reality.

The Genius in the Patent Office

Albert Einstein (1879–1955) was born in the city of Ulm in southern Germany. His father was a businessman who struggled to make his mechanical workshop turn a profit, while his mother was a well-educated woman with a passion for playing the piano. Albert may have inherited a fondness for gadgets and machines from his father: when he was four or five, his father showed him a magnetic compass; we can imagine Albert's eyes brightening as he pondered the invisible force that seemed to guide the device.

Never inspired by schoolwork, Einstein probably learned as much from the popular science books brought home by a family friend. On one occasion, Einstein was given an algebra book; he quickly whizzed through all the problems, even working out his own proof for the famous Pythagorean Theorem. Like Galileo before him, he was enchanted by the certainty of geometry and mathematics.

Landing a job, however, proved challenging for Einstein. He muddled through a series of temporary teaching positions, boosting his meager income through private tutoring. His personal life, too, was under strain. Both sets of parents disapproved of his relationship with Mileva Maric, a physics student. Einstein, unable to support a wife, delayed marriage. The two were still living apart when Mileva gave birth to their first child; the girl was given away.

In 1902, Einstein finally landed a job as a technical examiner in the Swiss patent office in Bern. Things now improved: within a year, he had enough confidence in his finances to ask Mileva to be his wife. The couple had two more children, both sons.

Einstein's days at the patent office have become something of a legend—the ultimate example of an unrecognized genius toiling away at a position far beneath his abilities. That view, to some extent, is quite justified: it gives us a romantic, but accurate, picture of the young scientist. We see him poring over reams of patent applications by day, then returning home to read up on the latest advances in theoretical physics by night. We picture him flipping the pages of physics journals with one hand, while gently rocking his son's cradle with the other. And yet, it would be a mistake to say that the patent job in any way slowed down

Albert Einstein in the patent office in Bern, Switzerland.

Einstein's progress. In fact, some scholars have argued exactly the opposite—that examining a large number of diverse technical proposals, scrutinizing them to determine which would work and which would fail, honed Einstein's mind in a most effective way. "The patent job agreed quite strikingly with his characteristic approach to his favorite problems in physics," says biographer Albrecht Fölsing. His daily duties as a patent examiner forced him to perform countless "thought experiments," Fölsing says, which would enhance Einstein's ability to reason by means of mental pictures and imagined machinery. The renowned physicist John Wheeler takes the argument a step farther: who else but a patent clerk, he asks, would have the mental equipment needed to see just what was wrong with contemporary physics?

> Day after day Einstein had to distill the central lesson out of objects of the greatest variety that man has power to invent. Who knows a more marvellous way to acquire a sense of what physics is and how it works?...Miracle? Would it not have been a greater miracle if anyone but a patent office clerk had discovered relativity? Who else could have distilled [the idea] from all the clutter of electromagnetism than someone whose job it was over and over each day to extract simplicity out of complexity?

Now, suddenly, Einstein's talents blossomed. This was a time of unmatched creativity, drive, and intellectual inventiveness. Never in the history of physics would one person accomplish so much so quickly. Though his ideas had been brewing for some time, they culminated in his "miracle year" of 1905. Einstein published six papers in the *Annalen der Physik* (*Annals of Physics*) that year—all of them paramount to the development of modern physics.

Einstein's first paper dealt with the way light interacts with metals; this paper, describing the *photoelectric effect*, was a key contribution to the new discipline known as quantum theory, and would eventually earn Einstein the Nobel Prize. (It would pave the way for such inventions as camcorders, digital cameras, and that light-sensor that keeps the elevator door from closing on you.) The second paper, on the dimensions of molecules, became one of the most-often cited in physics. The third and sixth papers dealt with "Brownian motion," the movement of microscopic particles when suspended in a fluid. It was his fourth and fifth papers, however, that shook the world of physics to its very foundation. Einstein, aged 25, had given us the beginnings of his *theory of relativity.*

Special Relativity

This principle ...has brought about a revolution in our physical picture of the world, which, in extent and depth, can only be compared to that produced by the introduction of the Copernican world system.

MAX PLANCK

Einstein began to formulate his theory of relativity as he pondered a very simple question: what would you see if you could catch up to a beam of light? According to the Newtonian picture, you'd see the beam of light as if it were standing still; your speed, relative to beam of light, would be zero. But Maxwell's picture contradicted this: his equations described light as a wave in motion, period. A light wave "standing still" simply held no meaning. Einstein's paper of June 30, 1905, "On the Electrodynamics of Moving Bodies," finally resolved the dilemma. (Because this paper dealt only with objects moving at constant speeds, it became known as the special theory of relativity, or *special relativity* for short. Later we'll see how Einstein expanded his theory to create the more complete *general relativity*.)

Einstein's solution was simple and yet surprising. The constant value of the speed of light is not some by-product of the way light travels through the ether; in fact, Einstein declared, there is no need to presume the ether even exists. Instead, he argued, the constant speed of light is simply a basic property of the universe. In other words, he postulated that the speed of light was always the same, independent of the motion of the observer. He added a second postulate which was very simple but just as important: he said the laws of physics must be the same for any two observers, no matter how fast they're moving relative to one another. That's why catching up to a beam of light seemed nonsensical—it would have meant that Maxwell's equations were valid for a stationary observer but not for one moving at high speed. That would mean that some observers have a more privileged perspective than others—something that even Galileo had realized was absurd.

Einstein's view of relative motion, based on his two postulates of special relativity, leads to some startling new effects. To get a clearer picture of this, let's go back to our train example. Instead of throwing a baseball, suppose you shine a flashlight in the direction of the train's motion. On board the train, we measure the light as traveling at the usual 300,000 km per second. What would an observer on the ground see? In the old Newtonian picture, the light would be given a "boost," that is, we would add the speed of the train to the speed of the light. That seems like a reasonable guess—but Einstein showed that it is wrong. It doesn't matter how much we try to boost the speed of light; it always travels at 300,000 km per second.

Of course, the speed of light is so large compared to the speed of a typical train that adding the latter would hardly make a difference, even in the old Newtonian picture. But suppose our train could reach the break-neck speed of 150,000 km per second—half the speed of light. A Newtonian would guess that, as seen from the ground, the light would have a speed equal to the sum of those two figures—450,000 km per second. But no—the light streaming out of the flashlight would *still* have its usual speed of 300,000 km per second. *You will always measure a beam of light as traveling at 300,000 km per second, no matter how fast or slow you are moving.* This means it is impossible to surpass the speed of light—it is, in effect, the ultimate speed limit in our universe.

The constant speed of light, however, is just the beginning. Remember, speed is always measured in terms of distance divided by time, as in "kilometers per hour." If the speed of light isn't affected by motion, then *space and time must be.* This was Einstein's startling prediction. In the old picture, described by Galileo and Newton, speeds were relative while space and time were constant. In the new picture, the only thing that is constant is the speed of light. Therefore, *space and time must be relative.*

Let's go back to our example of the moving train and see what happens to time measurements as the speed of the train increases. We need two people, one on board the train and the other on the ground. Let's call the passenger Alice and the ground observer Bernice. Since light moves at a constant speed, Alice can use a beam of light to build a simple clock on the train. Imagine a pulse of light traveling up and down between two mirrors attached to the floor and ceiling of the train (see (a) in the train illustration). One up-and-down cycle is one "tick" of the clock.

If the train isn't moving, Alice and Bernice both see the pulse of light doing exactly the same thing. The clock runs at the same rate for both of them. But suppose the train is moving with a speed close to the speed of light. Alice, moving along with the mirrors and the light beam, still sees the clock ticking at the old rate. But as seen from the ground, the picture is quite different:

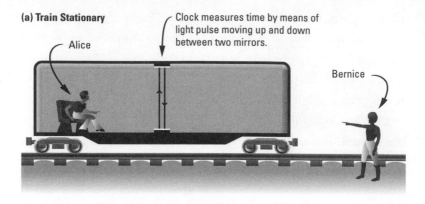

(a) Train Stationary

Alice

Clock measures time by means of
light pulse moving up and down
between two mirrors.

Bernice

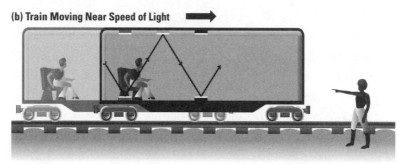

(b) Train Moving Near Speed of Light

(a) Train stationary: Alice (on board the train) has a clock that measures time by means of a light pulse moving up and down between two mirrors. While the train is stationary, both Alice and Bernice (on the ground) measure the same interval during each cycle of the clock.

(b) Train moving near speed of light: Bernice now sees the light pulse trace out a diagonal path. Since the speed of the light pulse is constant, she measures a greater interval during each cycle of the clock: the clock is "running slow." Bernice also sees the train (and everything in it) become shorter—but only in the direction of the train's motion.

Bernice now sees the pulse moving along a series of diagonal paths (see (b) in the illustration). Clearly, the pulse of light has to cover a greater distance between "ticks." Yet according to Einstein, the speed of the light pulse is the same. That means the time between each "tick" must increase. So Bernice says that Alice's clock is running slow. Alice, however, comes to the opposite conclusion: if there's an identical mirror clock on the ground, she would see *those* light pulses traveling along a diagonal route, and conclude that the clock on the ground is running slow! And, incredibly, they are both right; neither has a greater claim to the "correct" time.

This effect of slowing clocks—physicists call it *time dilation*—is just one of the bizarre results of special relativity. In our train example, Einstein's theory also predicts that the observer on the ground sees the length of the train, and everything in it, decrease—an effect known as *length contraction*. (This occurs only in the direction of motion, so Bernice sees Alice becoming thinner while her height and width remain the same.) Again, the situation is symmetric: Alice says that it's Bernice who is getting thinner!

Einstein also showed that the mass of an object moving at close to the speed of light, as seen by an outside observer, increases. This *mass increase* is harder to demonstrate with a simple diagram, but it is just as important as time dilation and length contraction. It explains, among other things, why the speed of light serves as the ultimate speed limit in the universe. Suppose you're in a spaceship, approaching the speed of light. You think, "I'll just step on the accelerator a little harder and I'll pass that pesky speed limit, no problem." But it won't work: to make your craft move faster, you have to use energy; the more massive the spaceship, the more energy you need. And, thanks to Einstein's special relativity, the mass keeps increasing, so you need more and more energy to further boost the speed. And you'll never quite make it. If you reached the speed of light, your spaceship, as seen by an outside observer, would have an infinite mass and it would have taken you an infinite amount of energy to get there. Flying at less than the speed of light, as physicists like to joke, isn't just a good idea—it's the law.

Einstein's description of space and time is shockingly counterintuitive—but only because our intuition comes from living in a world where speeds are always tiny compared with the speed of light. In everyday life, the effects of special relativity are very small. As an example, an average-size car moving at 80 km/h would be reduced in length by about the diameter of one atomic nucleus—far too small to notice. Even the astronauts who flew on the Apollo missions to the moon reached speeds of only 40,000 km/h—less than one hundredth of one per cent of the speed of light.

If a rocket were to approach the speed of light, however, the relativistic effects would soon become obvious. At half the speed of light, for example, it would appear about 14 per cent shorter. And at nine-tenths the speed of light, it would be reduced to less than half its normal length. (Again, an observer inside the rocket would see nothing unusual within the ship—but would see objects *outside* the craft reduced in length.) Clocks would slow and the rocket's mass would increase by the same proportion.

Even simple problems like adding up speeds have to be recast in the wake of relativity. Recall our example of the baseball thrower on board the train: when the ball was thrown, we added the speed of the ball to the speed of the train; the result was the speed that someone on the ground would measure. With high speeds—that is, speeds close to the speed of light—that simple formula no longer works. Einstein found the proper equation, and it supports the notion of the speed of light as the ultimate speed in our universe. No matter how high the two speeds that we add together are, the final sum is never greater than the speed of light.

Many people, even today, are disturbed by Einstein's theory. After all, it deals quite a blow to common sense. Some people try to console themselves by saying that relativity may describe how space and time would *appear*, but surely it doesn't tell us how they *really are*. Those clocks may appear to run slower, and those yardsticks may appear to shrink, skeptics argue—but it's just an illusion, right? Wrong. Although a quarter-century elapsed between the publication of Einstein's paper and the first direct tests of time dilation, length contraction, and mass increase, when the experiments were carried out, the results precisely matched the predictions of special relativity. In an accelerator, for example, protons can be sped up to such high speeds that their mass increases by a factor of more than 400. In 1971, relativity was tested by flying incredibly accurate atomic clocks around the world on board commercial jetliners. Afterwards, when the clocks were compared to identical atomic clocks on the ground, they were off by exactly the amount predicted by special relativity. Every experiment ever conducted to test the theory has confirmed Einstein's predictions.

The example with Alice and Bernice used a clock made from light and mirrors because it gave the clearest picture of the time dilation effect; however, because of the way relativity affects motion, any regularly repeating physical system—in other words, any clock—is affected in exactly the same way. Even our "biological clocks" are altered when we move at relativistic speeds. The best way of summarizing the effect is this: *It is time and space themselves that are distorted by relativity; our clocks and yardsticks are just along for the ride.*

Einstein's paper of June 1905 was not his last word on relativity. On September 28, the *Annalen der Physik* published his three-page follow-up, dealing with the relationship between mass and energy. Einstein showed that the two were, in fact, interchangeable, linked via $E = mc^2$, now the most famous equation in the world. (Here E stands for energy, m is mass, and c—from the Latin word *celeritas*, or swiftness, denotes the speed of light.) Einstein showed that not only does a moving body possess energy (the idea of "kinetic energy" goes back to Newton) but even a body at rest contains a certain amount of energy, "hidden" in its mass. The amount of this hidden energy, known as a body's *rest energy*, can be worked out using Einstein's equation. Because the speed of light is so large, even a small mass has a very large rest energy. Contrary to popular belief, any conversion of mass into energy—

in ordinary chemical reactions as well as in nuclear reactions—releases some of this hidden rest energy, but in most cases the effect is tiny and goes unnoticed. In nuclear power plants, however, a comparatively large amount of mass is converted into energy to produce electricity; in a nuclear explosion, the release is nearly instantaneous. The process can also work in reverse: in accelerators, for example, energy is converted into exotic particles of matter.

Einstein's equation is so well known that it has the reputation of some kind of "ultimate" formula. While $E = mc^2$ describes a great variety of phenomena and does give a unified picture of matter and energy, it is only a small step toward an overall unification in physics. It was not Einstein's greatest breakthrough, nor is it even the most important aspect of special relativity. (The theory's unified picture of space and time is surely just as vital.) Yet $E = mc^2$ is nothing to sneeze at. It governs the nuclear furnace that powers our sun and every other star in the cosmos, as well as the supernova explosions that spread dozens of chemical elements—including those needed for life—across the universe. It deserves its fame.

Einstein never described his work on special relativity as revolutionary. In his eyes, it was merely an extension of the work on electromagnetism begun by Faraday, Maxwell, and Hertz. And indeed it was—though it was the kind of extension that no ordinary thinker would have discovered. The originality of Einstein's work can be seen in that first paper on relativity in the *Annalen der Physik*. It runs 31 pages but contains no references—that is, he didn't need to cite any earlier work in order to lay out his ideas. For those who could appreciate Einstein's vision, relativity was a compelling, even beautiful work. As biographer Fölsing puts it, the early support for special relativity "was due not to conclusive experiments but to the fact that most physicists responded to its axiomatic, fundamental character and its beauty." Evaluating the theory's appeal many decades later and in rather more poetic terms, science writer Horace Judson declares that "No poem, no play, no piece of music written since then comes near the theory of relativity in its power, as one strains to apprehend it, to make the mind tremble with delight."

General Relativity

I cannot find time to write because I am occupied with truly great things. Day and night I rack my brain in an effort to penetrate more deeply into ...the fundamental problems of physics.
EINSTEIN, IN A LETTER TO HIS COUSIN ELSA

Einstein wasn't satisfied with special relativity. It was incomplete, he realized, because it dealt only with certain kinds of motion (specifically, motion at constant speeds) and it completely ignored gravity. For two years, the path to a broader theory eluded him. Then, in 1907, he had one of those "Eureka" moments. Using only his imagination, he discerned a deep connection between acceleration and gravity. Fifteen years later, in a lecture at Kyoto University in Japan, Einstein described his insight:

> The breakthrough came suddenly one day. I was sitting at my chair in the patent office in Bern. Suddenly a thought struck me: If a man falls freely he would not feel his weight. I was taken aback. This simple thought experiment made a deep impression on me. This led me to the theory of gravity.

Einstein's mental leap had led him to a link between acceleration—the motion of someone who "falls freely"—and the concept of gravity, the force that gives us the sensation of "weight." Every time you ride in an elevator, you can feel this connection: as the elevator starts upward—that is, as it briefly accelerates upward—your weight seems to momentarily increase. When it begins its downward journey, for an instant you feel lighter. We have similar sensations in cars, trains, or airplanes when they suddenly increase or decrease their speed. Today, pilots and astronauts even speak of the "g-forces" they encounter as they accelerate, where "one g" is equal to the force of gravity at the earth's surface, "two g's" is twice that force, and so on; the language itself incorporates this connection between accelerated motion and gravity. Perhaps the link seems obvious to us today—but it took the mind of Einstein not only to observe this connection but to work out its mathematical consequences. Through the power of pure thought, Einstein had found the key to unlocking the puzzle of gravity. It was "the luckiest idea of my life," he later said.

The link between gravity and acceleration, however, was just the beginning. Einstein still needed to find the mathematical laws that would describe how objects move in response to a gravitational field. (His effort parallels that of Newton in the 1660s, who first saw the connection between the falling apple and the orbital motion of the moon, and only then deduced the inverse-square law of gravity.) Einstein's task required herculean powers of analysis and, especially, mathematical insight.

In order to investigate his ideas about gravity, Einstein needed sophisticated new mathematical tools. The relevant equations and formulas had been invented only a few decades earlier by the German mathematician G. F. Bernhard Riemann (1826–66). That system, known as Riemannian geometry, describes the behavior of "curved" space—a far more flexible framework than the geometry Einstein had learned in school, which dealt with only "flat" space. (Riemann died of tuberculosis at 39, never learning of the rich harvest that Einstein would reap with the equations he had given the world.)

For nearly a decade, Einstein pondered the problem of gravity and the geometry of curved space. In the meantime, recognition of his talents gradually led to better-paying jobs. He resigned his post at the patent office in 1909, and took teaching positions in Zurich and later in Prague. In 1913 he was elected to the prestigious Prussian Academy of Science and offered a professorship in Berlin. He moved there with his wife, Mileva, and their two sons, though the couple soon separated. Through all of this, he toiled day and night over his equations. "In all my life I have labored not nearly as hard," Einstein said to a colleague. "I have become imbued with a great respect for mathematics, the subtler part of which I had in my simple-mindedness regarded as pure luxury until now. Compared with this problem, the original relativity is child's play."

Finally, in the fall of 1915, Einstein had fit all the pieces of the gravitational puzzle into place. He had worked out a broader framework that built on the earlier special relativity but which now encompassed accelerated motion and gravity. The new work became known as the general theory of relativity, or general relativity. By December of that year, Einstein was beaming: "The theory is beautiful beyond comparison," he declared.

We won't worry about the mathematical details of general relativity—even today, they are rarely taught to physics students below the graduate level. But the essence of the theory is actually quite easy to picture. For Newton, gravity was a force acting over a distance; for Einstein, it was a warping or curving of space itself. As an analogy, imagine a large, thick sheet of rubber. Roll a marble across the sheet and it moves in a straight line (see (a) in the illustration on page 107). Now imagine a bowling ball resting at the center of the sheet. Just by sitting there, it warps or distorts the rubber in its immediate vicinity. Roll a marble in the same direction across the sheet again, and its path curves because of the warping (see (b) in the illustration). This, Einstein showed, is how we should picture gravity. It is through such warping that the sun holds the earth in its orbit and the earth similarly holds the moon in its gravitational grip. In Einstein's world, matter distorts the very fabric of the universe. That distortion is what we experience as gravity.

Einstein's picture of the universe differs substantially from Newton's only in the neighborhood of intense gravitational fields; here in our solar system, that difference is rarely seen. There is one place, however, where the discrepancy was large enough to draw the attention of astronomers. Since the mid-1800s, they had noticed that Mercury didn't seem to obey Newton's laws precisely. As we saw earlier, Newton's law of gravity said that a planet's orbit should be elliptical. But the tiny, innermost planet, which makes one revolution around the sun every 88 days, didn't trace out a perfect ellipse. Rather, the planet's orbit shifted by a small amount each time it rounded the sun, gradually tracing out a sort of daisy-petal pattern. The effect—astronomers call it *precession*—is very slight; it amounts to less than a hundredth of a degree per century. Yet it defied a Newtonian explanation. Einstein's equations for general relativity finally explained Mercury's peculiar orbit; the amount of precession was exactly what Einstein's theory predicted. "For some days I was beyond myself with excitement," Einstein wrote to one colleague; the result "fills me with great satisfaction," he told another. His theory was not just idle mathematical speculation; it explained at least one long-standing problem in astronomy. He was now ready to tell the world about general relativity.

In November 1915, Einstein presented his theory in a series of lectures to the Prussian Academy in Berlin; the following year he set it down in writing in the *Annalen der Physik*. Those who could

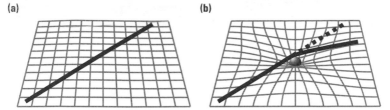

(a) A marble rolling along a flat rubber sheet travels in a straight line—just as light travels in straight lines through "flat" space.

(b) But massive objects distort the space around them, just as a bowling ball warps the rubber sheet. In this "curved" space, a rolling marble is deflected—just as light rays from a distant star are deflected as they pass our sun.

appreciate Einstein's achievement were in awe. As biographer Fölsing writes, "Einstein had attained a new level in the contemplation of nature, one that has been regarded as a model ever since but has never been repeated."

Just how original was general relativity? Historians have often speculated that had Einstein not developed his earlier theory of special relativity, one or more of his contemporaries would likely have found it. But on general relativity, the verdict is unanimous: this was "Einstein's baby," the work of a solitary genius who saw more deeply into the structure of the cosmos than anyone else of his time.

By explaining the anomaly in Mercury's orbit, general relativity was off to a good start. Explaining a phenomenon that was already known, however, is one thing; what the skeptics demanded was a prediction of some *new* phenomenon—followed by observations that would either support or refute the theory. (Remember, while working out general relativity, Einstein knew it had to account for Mercury's orbit—in other words, he knew the answer he was trying to get. As anyone who's struggled through Physics 101 knows, getting the right numbers is always much easier when the answers are provided at the back of the textbook.)

Einstein and the Eclipse

A more dramatic test of general relativity would come before the end of the decade. According to the theory, the sun should warp the space around it enough to bend the path of starlight that passes nearby. In other words, when the sun passes in front of a star—as seen from the earth—the star's position should appear to shift slightly. Normally, the sun is much too bright to allow such an observation; during an eclipse, however, the sun is blocked from view and the experiment can be performed. That's easier said than done, however. Total solar eclipses occur only a few times each decade and are visible only from a narrow strip of land somewhere on the earth's surface. Bad weather hampered an attempt to study an eclipse in South America in 1912; the First World War prevented an expedition to the Black Sea for the eclipse of 1914.

Finally, on May 29, 1919, a solar eclipse visible from the South Atlantic gave scientists their first chance to test Einstein's prediction. Two teams of British astronomers were dispatched, one to the central African coast and the other to Brazil. Developing the photographic plates and measuring the positions of the tiny stellar images took several months. Finally, on November 6, a joint meeting of the Royal Society of London and the Royal Astronomical Society was held to announce the results. The verdict: Einstein was right; the deflection was exactly the amount he had predicted. Though hesitant to disparage Newton—this was, after all, England—the scientists agreed that general relativity represented a new and more complete understanding of gravity. The president of the Royal Society, J.J. Thompson, called it "one of the highest achievements of human thought."

Back in Berlin, Einstein was calm and confident, never having doubted the outcome. When a student asked him what he would have done had the eclipse measurements not confirmed his theory, he replied: "In that case, I would have to feel sorry for God, because the theory is correct."

When the media picked up on the London announcement, Einstein was suddenly catapulted onto the world stage; he would remain a celebrity—albeit a reluctant one—for the rest of his life. Though already famous in German-speaking countries, Einstein reached superstar status with the coverage of the London meeting in *The Times* of November 7, 1919. "REVOLUTION IN SCIENCE," the paper declared in its headline; "NEWTONIAN IDEAS OVERTHROWN." Three days later, the *New York Times* seemed even more excited: "LIGHTS ALL ASKEW IN THE HEAVENS" and "MEN OF SCIENCE MORE OR LESS AGOG OVER RESULTS OF ECLIPSE OBSERVATIONS." Einstein's equations were later posted on the windows of a London department store, though the numbers and symbols probably meant little to the average shopper.

Einstein became the talk of the town—not just in Berlin and London, but around the world. "At present, every coachman and every waiter argues about whether or not the relativity theory is correct," Einstein wrote to a friend in 1920. Sightseers came to Berlin to sit in on his lectures at the university—unable to comprehend the material, however, they frequently left after the first few minutes. From California to Japan, crowds now routinely greeted Einstein wherever he traveled.

By now a Nobel Prize seemed inevitable, though this presented a dilemma for the Royal Swedish Academy of Sciences, which oversees the awards. As biographer Fölsing points out, the Academy needed to recognize Einstein—but relativity was still too controversial. They solved the problem by honoring him instead for his investigation of the photoelectric effect, published back in 1905. The move, writes Fölsing, "saved the Swedish Academy from a delicate situation; by then the academy needed Einstein as a laureate more than he needed the prize." The money, of course, was still relevant; Einstein gave it to his ex-wife as part of their divorce settlement.

The public was anxious to read about this new superstar of science. Whether they understood his work, however, is another matter. In London, the editors of *The Times* confessed that they "cannot profess to follow the details and implications of the new theory with complete certainty," though even the leading physicists of the day "find no little difficulty in making their meaning clear." Within a year of the eclipse announcement, more than 100 books on relativity appeared. The magazine *Scientific American* offered a $5000 prize for best popular account of relativity; Einstein quipped that he was the only one not seeking the prize. Einstein did write articles on relativity for the world's leading newspapers, as well as a popular book, *Relativity: The Special and the General Theory*, still in print after more than 85 years. Gradually, the public response to Einstein went through a transformation. It was no longer the theory itself, but the fact that it seemed to defy comprehension that gave relativity its near-mystical appeal.

The Later Years: Einstein's "Lonely Old Song"

Never comfortable in the spotlight, Einstein tried to ignore the media circus that was now a permanent part of his life. He married his cousin Elsa in 1919; she remained by his side until her death 17 years later. When the Nazis came to power in 1933, the couple left Berlin and Einstein renounced his German citizenship. Einstein moved to the United States, where he took a position at the newly created Institute for Advanced Study in Princeton, New Jersey. He was politically active throughout his life, supporting Israel and becoming a figurehead of the new pacifist movement. Despite his hatred of war, Einstein wrote to President Roosevelt urging the U.S. to develop an atomic bomb before Germany. (Einstein himself, however, was never directly involved in the Manhattan Project.)

With general relativity behind him, Einstein again began to look at the deepest problems in physics—including the quest for a unified theory. He spent his final decades trying to unify gravity with electromagnetism, completely ignoring the fact that mainstream physics was proceeding in quite different directions. Einstein had never been comfortable with the other great development of twentieth-century physics—quantum theory—even though by the middle of the century it had paved the way for dozens of spectacular developments in atomic and nuclear physics.

During his final years, colleagues saw Einstein as something of a museum piece, an old man with old ideas. Einstein knew he was becoming isolated from mainstream physics, but preferred, as he put it, to sing his "lonely old song" than to follow the latest trends. As biographer Fölsing puts it: "Even devoted admirers of Einstein would not dispute that the progress of physics would not have suffered unduly if the indisputably greatest scientist among them had spent the final three decades of his life—roughly from 1926 on—sailing." Einstein died in the spring of 1955; his body was cremated after a simple ceremony.

Timewarps, Black Holes,
and Gravitational Waves

During Einstein's lifetime, general relativity passed two tests. It explained the odd behavior of Mercury's orbit, and it predicted the bending of starlight by the sun, confirmed with the 1919 solar eclipse results. Some 40 years would pass before the more difficult predictions of the theory could be put to the test.

One of the more mind-bending predictions of general relativity is that matter warps time as well as space. The effect is harder to visualize, but it means that a clock in a strong gravitational field will run slower than an identical clock in a weak field. For example, a clock at the base of a mountain will run slightly slower than a clock at the peak, simply because the lower clock feels the earth's gravity more intensely. The effect is small and was impossible to verify in Einstein's day, but has since been confirmed by many different tests.

Perhaps the most remarkable prediction of general relativity is the existence of exotic realms in which space and time are warped so severely that they become, quite literally, cut off from the rest of the universe. Thanks to science fiction writers and innumerable episodes of *Star Trek*, everybody has heard of them. I'm speaking, of course, about *black holes*. According to general relativity, black holes can form wherever a large amount of mass is confined in a very small space, for example, when a massive star collapses after exhausting its nuclear fuel supply. If the star is big enough, the resulting collapsed core will have such an intense gravitational field that nothing—not even light—can escape.

By definition, such objects are "black" and cannot be seen directly. Nonetheless, there is now a nearly rock-solid case for the existence of black holes. The first studies involved the motion of normal stars seen to be orbiting massive but invisible companions. More recently, astronomers have detected what appears to be the accelerated swirling of matter as it is "swallowed" by a black hole; the in-falling matter can be detected through the X-rays it given off just before it disappears. There has also been mounting evidence that so-called "supermassive black holes" lurk in the cores of most galaxies, including our own Milky Way. Today, many physicists suspect that giant black holes may be among the most ancient and important objects in the universe, likely guiding the evolution of the galaxies that now harbor them.

Another novel prediction of general relativity is only now being tested. According to Einstein's theory, the motion of a massive object should emit a kind of radiation. Just as Maxwell showed that accelerating electrical charges emit electromagnetic waves, Einstein showed that accelerating masses should emit *gravitational waves*. These waves should alternately stretch and shrink space itself as they pass by. Gravitational waves are extremely elusive, however, because gravity is such a weak force; so far, they have not been directly measured. Deep in the cosmos, however, intense gravitational waves—perhaps large enough to detect from the earth—should be emanating from regions where massive objects are throwing their weight around: rapidly spinning pairs of black holes, for example, or collisions between ultra-dense objects such as white dwarfs or neutron stars. A number of sophisticated detectors, which scientists hope will finally glimpse such waves, will soon begin operation.

However, there is already strong evidence that gravitational waves exist. Beginning in the mid-1970s, U.S. astrophysicists Joseph Taylor and Russell Hulse made careful observations of a system known as a binary pulsar—a pair of small, dense stars revolving rapidly around one another. According to general relativity, the pair should radiate energy in the form of gravitational waves. And this energy loss, in turn, should cause the two stars to slow down in their orbits. Measurements showed that the pair were indeed losing energy—and at precisely the rate predicted by the theory. Taylor and Hulse shared the Nobel Prize in 1993 for their work.

Relativity and the Cosmos

Alvy Singer, age nine: The Universe is expanding. The universe is everything, and if it's expanding, someday it will break apart and that will be the end of everything.

Alvy's mother: What is that your business! You're here in Brooklyn. Brooklyn is not expanding!

FROM WOODY ALLEN'S *ANNIE HALL*

There is one more object governed by general relativity that we have not yet discussed: the universe itself. Because gravity is the dominant force in the cosmos, the study of *cosmology*—the effort to understand the origin and evolution of the universe—is ultimately founded on the framework of general relativity.

Einstein realized the importance of general relativity for cosmology soon after he worked out his equations. At that time, however, so little was known about the structure of the universe that he let his cosmological prejudices lead him astray. The equations of relativity did not allow a "static" universe; instead, the theory said that the universe must either be expanding or shrinking. This idea made no sense to Einstein, so he introduced a "fudge factor" into his equations—he called it the *cosmological constant*—in order to keep the universe stable. In 1929, however, U.S. astronomer Edwin Hubble made a remarkable discovery. By studying the light from distant galaxies, he found that they were receding; in fact, the more distant the galaxy, the faster it was speeding away. The universe was indeed expanding! When Einstein learned of the discovery, he immediately regretted introducing the cosmological constant, calling it the "greatest blunder" of his career.

If the universe is expanding, it means that at some time in the past, those galaxies must have been much closer to each other. In fact, if we go far enough back in time, we reach a phase in which the universe was much smaller and much hotter than today. Scientists concluded that the universe began in an ultra-dense fireball and has been expanding and cooling ever since. This is known as the "big bang" model of the universe. Further evidence for the big bang came in 1965, when astronomers discovered that the universe was awash in microwave radiation produced in that initial fireball—the faint "echo" of the big bang itself. The big bang model is now the cornerstone of modern cosmology. Astronomers believe that the initial explosion occurred about 15 billion years ago. They're now busy pinning down the other parameters of the model, such as the universe's density and its expansion rate.

In an ironic twist, however, Einstein's cosmological constant may be making a comeback. Careful observations of remote galaxies show that they are not only speeding away from the earth, they are also accelerating. While gravity draws matter together, a mysterious force is having the opposite effect, pushing galaxies apart. So far, the nature of this force is unknown, though one possibility is that Einstein's "fudge" was in fact correct—that there really is a cosmological constant, and that it is just as important as gravity in shaping the evolution of the universe.

Einstein didn't just seek out a unified theory of physics; he personified that search. No one can remain unmoved by the story of Einstein's struggle to find a grand, comprehensive theory that would unite all of physics—a search that ultimately left him frustrated. Coming up with various laws describing various phenomena meant nothing to him. What he wanted, what he desperately sought, was a single, unified theory. His work on relativity, he wrote, was only the first small step, advancing "nearer to the grand aim of all science, which is to cover the greatest possible number of empirical facts by logical deduction from the smallest possible number of hypotheses or axioms."

With special relativity, Einstein integrated time and space, and showed that matter and energy are two sides of the same coin. We may be tempted to put his famous equation, $E = mc^2$, on our T-shirt—but this would focus too much attention on too narrow an achievement. After all, special relativity, where the equation originated, was just the beginning. It was general relativity, Einstein's masterwork, that showed the all-important link between gravity and the geometry of space. So it is the equations of general relativity that we place on our T-shirt; if a visual aid is needed to accompany the math, we can add the metaphor of the bowling ball on the rubber sheet from page 107; it is as close as we can come to embracing Einstein's theory without becoming mired in mathematics.

Over the last 85 years, general relativity has become one of the great pillars of modern physics. Golf balls and galaxies, black holes and the big bang—all are governed by its far-reaching equations. As we'll see in the next two chapters, however, making general relativity consistent with the other great pillar of modern physics—quantum theory—has become the foremost challenge in today's quest for unification.

Quantum Theory and Modern Physics

Things Get Even Weirder

Anyone who is not shocked by quantum theory has not understood it.

NIELS BOHR

To make a discovery is not necessarily the same as to understand a discovery.

ABRAHAM PAIS

In the last chapter, I described Einstein's theory of relativity as if it were the greatest challenge to common sense that twentieth-century physics had come up with. It is not. That honor goes to quantum theory.

Relativity, as we have seen, was needed because of a subtle but troublesome failing of Newtonian mechanics. In certain kinds of experiments, Newton's laws failed to give the right answers—specifically, experiments that involved speeds close to the speed of light (special relativity) or strong gravitational fields (general relativity). Toward the end of the nineteenth century, however, physicists were discovering another realm in which Newton's picture of the world was inadequate—the realm of the atom. A new picture, founded on quantum theory, would eventually replace Newtonian mechanics. Like relativity, quantum theory was a radical departure from the prevailing description of the world. The old Newtonian picture (which became known as "classical" mechanics) described a predictable, mechanistic universe; if you knew what a physical system was doing at any moment, you could, in principle, work out what it would do next. That predictability, however, would soon give way to a shadowy world of probabilities and paradoxes. Quantum theory was so radical that even Albert Einstein—one of its founders—could not accept it. But unlike relativity, which was largely the work of one man, quantum theory was born from the insights of nearly a dozen great thinkers over a period of nearly 30 years. We begin our exploration of the quantum world by taking a journey into the atom.

A Brief History of the Atom, Part One

The idea of the atom was not new, of course; we encountered the idea back in the first chapter in the bold speculations of Leucippus and Democritus. These daring Greeks, however, were far ahead of their time. Atoms really do exist, it turned out—but the tools needed to probe them directly didn't arrive until the Scientific Revolution was well underway.

Modern atomic theory began with the British scientist John Dalton (1766–1844), who explained the properties of certain chemicals by assuming that they were composed of different varieties of atoms. "All bodies of sensible magnitude," he wrote, "are constituted of a vast number of extremely small particles, or atoms of matter, bound together by a force of attraction...." Scientists soon realized that although there were many different kinds of atoms, they always combined with each other in certain exact proportions. (Water, for example, always requires two atoms of hydrogen for each atom of oxygen.)

During the nineteenth century, dozens of new kinds of atoms were discovered. Keeping the Greek terminology, substances made from a single type of atom were called *elements*. Though these elements possessed a range of atomic sizes and chemical properties, once again a pattern was concealed within the diversity. In 1869, the Russian scientist Dmitri Mendeleyev (1834–1907) published a theory describing the regularity of atomic properties—the now-ubiquitous *periodic table* that graces chemistry classrooms around the world.

In a flurry of activity around the turn of the twentieth century, physicists began to probe the nature of the atom itself. A high-energy form of electromagnetic radiation—X-rays—was discovered in 1895; the radiation emitted by unstable atoms—radioactivity—in 1896. The following year, British physicist J.J. Thompson (1856–1940) discovered the electron. By this time, it was clear the atom was not truly indivisible. Instead, atoms appeared to be made up of certain electrically charged particles, that is, particles that carry either a positive or a negative charge. The electron was clearly one of those components. Because electrons carry a negative charge, however, something else within the atom must be carrying a positive charge; after all, most substances were seen to be electrically neutral.

But the actual structure of the atom—the arrangement of those charges—was unknown. Thompson suggested they were evenly spread throughout the atom; this was known as the "plum pudding" model (imagine plums representing the negatively charged electrons, spread evenly through the positively charged dough). It was a reasonable enough guess—but one which had not yet been tested by experiment. Enter the brilliant New Zealand-born scientist, Ernest Rutherford (1871–1937). In 1907, Rutherford, who had been working at McGill University in Montreal, moved to England to take up a professorship at the University of Manchester. It was there that he devised an ingenious experiment to probe the structure of atoms. He took heavy, positively charged particles known as alpha particles (they're what you get when you take a helium atom and strip away the electrons) and fired them at a thin sheet of gold foil. The speedy particles should have streamed right through the gold foil like hot knives through butter. Indeed, most of them did just that. But some of the alpha particles were deflected; a few even bounced straight back. "It was quite the most incredible event that has ever happened to me in my life," Rutherford said of the result. "It was almost as incredible as if you fired a fifteen-inch shell at a piece of tissue paper and it came back and hit you."

What caused the alpha particles to make a U-turn? The simplest explanation, Rutherford reasoned, was that the positive charge of the atom must be concentrated in one small area—providing a massive, positively charged "target" that repelled the positively charged alpha particles. Instead of being spread evenly throughout the atom, as Thompson had proposed, the atom's positive charge seemed to be found only in the center. Later experiments confirmed Rutherford's theory; today this cluster of matter at the heart of an atom is called the *nucleus* (plural *nuclei*). The individual particles that carry the positive charge are called *protons*. (The more slippery *neutron*—as heavy as a proton, but carrying no electrical charge—eluded discovery until the 1930s.)

Usually, the number of protons in an atom exactly matches the number of electrons, which Rutherford pictured as orbiting the nucleus in a manner analogous to the way the planets orbit the sun. And, just like in our solar system, most of the structure is empty space: experiments later showed that the radius of a typical nucleus is only one ten-thousandth that of its host atom. If we scaled an atom up to the size of Manhattan, the nucleus would be the size of a compact car in the middle of Central Park.

With this new model of the atom in mind—an atom consisting primarily of empty space, but with a tiny high-density "clump" in the centre—it's easy to explain the strange result Rutherford saw in his gold foil experiment. Most of the incoming particles would pass right through the foil, never striking the tiny nuclei. But the small number that happened to hit a nucleus dead-on would rebound—just as he observed.

Rutherford was already a Nobel Laureate by the time he conducted his groundbreaking work on atomic structure; he had won the 1908 prize for chemistry for his work on radioactivity. He received further honors, becoming a knight in 1914 and a baron in 1931. None of these English honors, however, could take the New Zealander out of Rutherford. On one occasion, during a formal luncheon, a bishop was stunned to hear that Rutherford's native South Island had a population of only 250,000, and reflected that more people lived in the English city of Stoke-on-Trent. That may be so, Rutherford conceded. "But let me tell you, Sir, that every single man in the South Island of New Zealand could eat up the whole population of Stoke-on-Trent, every day, before breakfast, and still be hungry."

Rutherford's picture of the atom, however, was clearly incomplete. For one thing, an atom that really functioned like a miniature solar system could not be stable. As we saw earlier, Maxwell showed that an accelerating charged particle, such as an electron orbiting a nucleus, emits radiation; as it does so, it loses energy. What, then, would prevent the negatively charged electrons from spiraling into the positively charged nucleus?

Another difficulty had bothered physicists even longer. They knew that hot objects radiate energy; the hotter the object, the brighter it glows. By the late nineteenth century, physicists had learned how to measure exactly how much light of different colors these objects produced—what they called the object's *spectrum*. (In the language of physics, the spectrum records the intensity of the radiation given off at different wavelengths.) But Maxwell's theory of electromagnetism gave wildly inaccurate predictions of those spectra. According to the old theory, as an object is heated, the most intense radiation should come at shorter and shorter wavelengths. When you switched on a stove, you should be blinded by a burst of ultraviolet radiation and X-rays. Obviously, this does not happen. The solution would come from an unlikely revolutionary—the German scientist Max Planck.

Max Planck and the Birth of the Quantum

Max Planck (1858–1947) hailed from a staunchly conservative background, "a family of ministers and lawyers—upright, dutiful, honest people," as one historian put it. His thesis advisor tried to steer him away from theoretical physics, saying that after Newton and Maxwell there was little work left to do. Fortunately, Planck ignored that advice; in 1889 he was awarded a prestigious position at the University of Berlin. He was already 42, with some 40 published papers behind him, when he turned his attention to the mystery of the radiation of hot objects.

On December 14, 1900, Planck presented his solution to a meeting of the German Physical Society in Berlin, in a paper titled, "On the Distribution of Energy in a Normal Spectrum." In what he described as "an act of desperation," he suggested that energy spilled out of hot objects not in a continuous stream but in discrete bundles of energy of a particular size. (A comparison is often made to automated bank machines, which dispense money only in multiples of, say, $20.) Planck called each of these energy bundles a *quantum* (plural *quanta*), from the Latin word for "how much."

Max Planck's investigation of the radiation emitted by hot objects triggered the quantum revolution.

How big is a quantum of energy? Quite small, as it turned out. It depends on a particle's frequency and on a number known as "Planck's constant," denoted by the symbol h. In the system of units used by physicists, in which mass is measured in kilograms, distance in meters, and time in seconds, the value of h works out to about 0.000000000000000000000000000000000066, or 6.6×10^{-34} (this shorthand, called scientific notation, is explained in the Note to the Reader). Because Planck's constant is such a small number, quantum effects go unnoticed in the everyday world. It's as if we're measuring the economic activity of a large company, seeing only multi-million-dollar revenues and expenses, but ignoring the individual banknotes that make up those transactions.

We know, of course, that money is "quantized"—it comes in bills and coins of fixed amounts—but why should energy follow such a seemingly arbitrary rule? Planck himself felt the quantum hypothesis was absurd; he readily admitted he had no idea what it actually signified. It did, however, give a simple solution to the problem at hand. Using the quantum hypothesis, Planck could precisely account for the spectra produced by hot bodies. But quantum theory would reach much farther. It soon led to a radical new picture of how matter and energy behave at the atomic level.

Einstein and Bohr: Quantum Theory Takes Shape

Let's take a closer look at those hot objects that Planck investigated. They emit radiation—but where does that radiation come from? Presumably, physicists said, from the motion of charged particles within the atoms themselves, specifically the electrons, thought to be in constant motion around the nuclei. (Maxwell's theory, again, predicts that accelerating, charged particles emit radiation in just such a fashion.) Now, thanks to Planck, it became clear that those whizzing electrons could have only specific energies—that is, that the energy of each electron must be quantized. Albert Einstein took the next step: if the energy of the electrons is quantized, then what about the light they give off? In his study of the photoelectric effect—his first 1905 paper—Einstein showed that light, too, comes in discrete bundles.

As we saw earlier, the nature of light had long puzzled physicists. By the end of the nineteenth century, however, there was overwhelming evidence that light was a *wave*. Now Einstein had shown that it could also be considered a *particle*, referred to as a photon. Einstein showed that the energy of a photon depends on its wavelength (or, equivalently, on its frequency) and on Planck's constant, h. Because h is so tiny, however, the energy carried by any one photon is extremely small. Indeed, if you switch on a 100-watt lightbulb and switch it off one second later, you've unleashed some 300 billion billion, or 3×10^{20}, individual photons.

The next step in the quantum revolution came with the work of the Danish physicist Niels Bohr (1885–1962). In 1913, Bohr, still a newcomer to the world of theoretical physics, applied the new quantum theory to the hydrogen atom. Hydrogen is the most abundant element in the universe and also the simplest; it consists of a single proton with a single electron circling around it. Bohr showed that the electron could only occupy orbits that had specific, discrete energy levels. (One might compare the electrons to people attending a concert in an auditorium, who may find themselves, say, in row 12 or row 13, but never row 12.5.) In this picture of the hydrogen atom, radiation is emitted only when the electron "jumps" from one energy level to another. The bigger the electron's jump, the greater the energy of the photon released.

Niels Bohr, pictured here with Einstein, applied the principles of quantum theory to the hydrogen atom. Einstein, though skeptical of the new theory, also contributed to the quantum revolution with his discovery of the photoelectric effect.

Bohr's proposal was radical. It completely contradicted the ideas of Newton and Maxwell—but, as physicists were discovering, the classical framework needed to be overhauled in order to explain the workings of the atom. The new theory immediately showed its strength. First, it solved the problem of atomic stability; second, it described the spectrum of hydrogen with incredible precision. Indeed, the theory's prediction of the locations of the various "peaks" in the hydrogen spectrum agreed with those seen in the laboratory to about one hundredth of one per cent. Later, the quantum hypothesis was applied to other elements with equal success—even allowing scientists to predict the properties of exotic chemicals not yet discovered. Awarded the Nobel Prize in 1922, Bohr is recognized alongside Planck and Einstein as one of the key players in the early development of quantum theory.

De Broglie, Heisenberg, and Schrödinger: The Fuzzy World of Quantum Mechanics

The development of quantum mechanics has been a sobering experience. It has shown where the limits of intuitive comprehension lie.

VICTOR GUILLEMIN

By 1920, quantum theory had shown a number of great successes, but physicists were still in the dark about what the theory actually meant. It didn't say how we should *picture* the behavior of electrons and photons and other bits of matter and energy. That would change during the 1920s, beginning with the work of a young Frenchman, Louis de Broglie (1892–1987). De Broglie was a history student, but he took an interest in science after being posted at the Eiffel Tower radio station in Paris during the First World War. He later earned a Ph.D. at the Sorbonne, where he remained a professor for nearly 40 years. De Broglie's greatest work, however, came early in his career. He knew about Einstein's discovery that light, long regarded as a wave, was now also seen as a particle. Perhaps electrons, thought to be solid particles, might behave like waves. He first described this idea of "matter waves" in his doctoral thesis of 1923. The notion was so bold that his examining committee didn't know what to make of it; only when they appealed to Einstein in Berlin, who gave the proposal a thumbs-up, did they award de Broglie his Ph.D.

De Broglie's proposal raised a number of perplexing questions. If a particle could behave like a wave, what was its wavelength? According to de Broglie, the wavelength could be worked out by dividing Planck's constant by the particle's *momentum* (a number you get by multiplying a particle's mass by its speed). Of course, Planck's constant is already a very small number—so the only way for the end result to be large is if the momentum is equally tiny. For large objects moving at everyday speeds, the size of these matter waves is negligible, so the wave nature of large (macroscopic) objects rarely reveals itself. In the atomic and subatomic realm, however, that size becomes significant, and the wave-like properties of matter cannot be ignored.

American Institute of Physics

One of the founders of quantum theory, Louis de Broglie discovered that particles of matter could also behave like waves.

By the late 1920s, the wave nature of electrons was already turning up in experiments. Physicists had bounced streams of electrons off dense crystals, in which atoms are tightly packed in a regular array. They observed "interference patterns"—the characteristic patterns you get when waves overlap with each other (similar to the water waves shown on page 84). De Broglie was right: a beam of electrons can behave as if it's made up of waves, and the size of those waves was exactly what his theory predicted. And yet, in other experiments, electrons behaved in the more "traditional" way—as little bits of matter.

De Broglie received the Nobel Prize in 1929 for his work in revealing this *wave-particle duality*, a kind of Jekyll-and-Hyde dualism in the subatomic world. Depending on the kind of experiment being performed, electrons—and indeed all subatomic particles—can display particle-like or wave-like properties. In an interesting historical quirk, one of the decisive electron-wave experiments was carried out by British physicist George Thompson—son of J.J. Thompson, who had discovered the electron back in 1897. The father won a Nobel Prize for showing that electrons were particles; his son won a Nobel for showing they were waves. Both were right.

So far, these advances may sound like a rather motley crew of unrelated—and rather puzzling—findings. True, quantum theory was successful—but no unifying principle held it all together. As physicist Wolfgang Pauli put it in 1925, "physics at the moment is again very muddled." Within a few years, however, a new mathematical framework for quantum theory took shape, with wave-particle duality at its heart. This new description of matter was developed by a number of physicists, led by the German Werner Heisenberg (1901–1976) and the Austrian Erwin Schrödinger (1887–1961). The new framework became known as *quantum mechanics*, to distinguish it from the earlier quantum theory from which it developed. (For our purposes, however, it is safe to use the two terms interchangeably.)

Heisenberg, poring over his equations, was struck by the consistency and elegance of the emerging mathematical picture: "I had the feeling that, through the surface of atomic phenomena, I was looking at a strangely beautiful interior, and felt almost giddy at the thought that now I had to probe this wealth of mathematical structures nature had so generously spread out before me." Schrödinger, meanwhile, set out to find the equation that governed the behavior of those matter waves that de Broglie had proposed. His great breakthrough is said to have come in December 1925, during a secret rendezvous with his mistress at a ski resort in the Swiss Alps. (His wife, we're told, was also cheating on *him.*) His relationships may have been turbulent, but his math was rock-solid. The wave equation at the heart of quantum mechanics is now known as the *Schrödinger equation.* Both Heisenberg and Schrödinger were duly recognized with Nobel Prizes, in 1932 and 1933 respectively.

Werner Heisenberg, discoverer of the "uncertainty principle."

The mathematics of quantum mechanics, while a great triumph for those trying to work out the properties of atoms and subatomic particles, led to a more and more outrageous description of nature. Indeed, wave-particle duality was just the beginning. Heisenberg discovered that a particle's position and its speed can never be known at the same time; the more precisely you know one, the less you know about the other. Trying to pin down both is a bit like grappling for the soap in the bathtub; just when you think it's in your grasp, it slips away. In the case of atomic or subatomic particles, however, the slipperiness is absolute. The effect is known as the *uncertainty principle.* It has nothing to do with the limitations of our measuring devices—it is simply a property of nature itself. And, once again, this quantum uncertainty never shows itself in the everyday world because Planck's constant, h, is so small. When a billiard ball rolls across the table, for example, the uncertainty is only about one part in 10^{27}, or one in a billion billion billion—far too small to have any noticeable effect—and you can safely use classical mechanics to line up your shot. On the subatomic scale, however,

Erwin Schrödinger worked out the wave equation that governs the atomic and subatomic world.

we have speeds and sizes that are on par with h, and quantum uncertainty dominates the scene.

Indeed, any kind of measurement in the quantum world is plagued with difficulty. It turns out that we can only calculate the *probability* of different results. In fact, there is no guarantee that the same experiment, carried out a second time, will give the same outcome. Quantum theory only allows us to compute the *average* of many such measurements—or, more precisely, the probability that each measurement will fall within some particular range of values. As an analogy, imagine you're a director of the Metropolitan Museum of Art in New York, and you're trying to predict how many visitors will enter the gallery next Sunday. There is no way to predict the exact number; the best you can do is look at the ticket stubs from the past month or the past year, and see on average how many visitors usually drop by on Sundays. Business owners of all stripes have learned to take such uncertainty in stride; the insurance industry, in fact, is founded on probabilities. But physics—at least physics as it was then known—was never thought to be subject to such unpredictability. Just the opposite, in fact: for more than two centuries, the successful application of Newton's mechanics had left scientists with little reason to believe the universe was anything but predictable.

Following the work of Heisenberg and Schrödinger, however, physicists were left with an all-pervasive uncertainty. One element of predictability did survive: one can still predict how those waves described by Schrödinger's equation evolve over time. But such calculations only yield the probability of some quantum event occurring. In the atomic world, at least, the new picture seemed to be inherently fuzzy. We saw earlier how Rutherford pictured the electron as orbiting the nucleus rather like a miniature solar system. Now we see that this is a much too mechanical view. A better way of describing an atom is to say that the electrons are "smeared out" in a sort of cloud that surrounds the nucleus.

But the weirdness of the quantum world doesn't end there. The theory, scientists discovered, implies a fundamental link between the observer and the thing being observed. Until we look at a particle, for example, it can literally have any speed and be located anywhere—in the jargon of particle physics, it can be in any *state*. Indeed, until we measure it, the particle can be in many states at once what's known as a *superposition* of states. When we actually make a measurement, however, we somehow force the particle to "collapse" into a single state. Quantum mechanics can only tell us that when we finally perform our measurement, some states are more likely to be observed than others.

Don't worry if all of this sounds very bizarre; it *is* bizarre. None of these concepts existed in classical mechanics. In fact, many of the scientists involved in developing quantum theory were deeply disturbed by its implications. That was particularly true of Einstein. Even though his own work was crucial in laying the foundations of the theory, he was never satisfied by the probabilities and fuzziness inherent in the new quantum picture. Indeed, he carried on a decades-long debate with several colleagues, arguing over the merits and supposed shortcomings of quantum mechanics. Referring to God as the "old one," Einstein expressed his dissatisfaction in a famous letter to Max Born in 1926: "The theory says a lot but does not really bring us any closer to the secret of the 'old one.' I, at any rate, am convinced that *He* is not playing at dice." Today, however, it seems that Einstein was wrong on this matter, and that the universe is indeed founded on quantum probabilities. Seventy years after Einstein's frustrated remark, Stephen Hawking told a lecture audience: "It seems that even God is bound by the uncertainty principle....All the evidence points to him being an inveterate gambler, who throws the dice on every possible occasion."

Schrödinger's Cat and Other Tales of "Quantum Weirdness"

Gone is the ideal of a universe whose course follows strict rules, a predetermined cosmos that unwinds itself like an unwinding clock. Gone is the ideal of the scientist who knows the absolute truth.

HANS REICHENBACH

Perhaps the most troubling aspect of quantum mechanics is the notion of superposition—the peculiar idea of a particle being in two states at once. While this "quantum weirdness" may not be a problem for subatomic particles, in certain situations it could spill over from the atomic world into the everyday world. In 1935, Schrödinger proposed a famous "thought experiment" in which he described just such a situation.

Schrödinger asked us to imagine a cat in a sealed box. Also in the box is a small amount of radioactive material, a Geiger counter, and a vial of poisonous gas. If one of the atoms of the radioactive material decays—a quantum event—it is detected by the Geiger counter, which in turn causes a hammer to break the vial, releasing the gas and killing the cat. (Remember, it's just a thought experiment!) Let's say that quantum theory gives a 50-50 chance that one of the radioactive atoms will decay over a one-hour period. After an hour has gone by, is the cat alive or dead? When you open the box, of course, you'll find out. At that

Physics Today

The famous "Schrödinger's cat" thought experiment. There's a 50-50 chance of a quantum event triggering the release of a poisonous gas, killing the cat. Until we make a measurement—that is, until we open the box and look inside—quantum mechanics says the cat is in a superposition of "living" and "dead" states; that is, the cat is alive and dead at the same time.

moment, the system is said to "collapse" into one specific state, revealing either a live cat or a dead cat. Until you open the box, however, the cat is said to be in a superposition of "living" and "dead" states—that is, the cat is alive and dead at the same time.

Perhaps we can imagine an electron or some other subatomic particle occupying two states at the same time, but what are we to make of a cat supposedly hovering in an alive-dead state? The implications of Schrödinger's cat have been endlessly debated—so much so that many physicists are tired of hearing about it. Stephen Hawking once said, "When I hear of Schrödinger's Cat, I reach for my gun." Yet it raises questions that have yet to produce satisfying answers. The dilemma of Schrödinger's cat forces us to search for the realm where the laws of physics "cross over" from the quantum to the classical. It asks us for an "interpretation" of quantum mechanics in the everyday world. A number of such interpretations have been put forward, none of them pleasing to everybody. The two most popular versions are the Copenhagen and the Many Worlds.

The Copenhagen Interpretation is named for the city where Niels Bohr, one of its primary advocates, lived—although Bohr never explicitly defined it. In this interpretation, quantities that have not been measured—like the cat's condition prior to opening the box—simply have no meaning. The only meaningful quantities are the results of measurements. (A similar viewpoint was expressed by the philosopher David Hume back in the eighteenth century.) This interpretation, however, raises the questions of exactly *who* is doing the measuring and what qualifies as a measurement. Is a conscious observer needed in order for the theory to make sense? A few scientists suggest that consciousness is indeed tied inexorably to the quantum world, and hope that some far-reaching physical theory will explain *both* how the mind works *and* quantum weirdness. (As you can imagine, though, this leads to even more questions. If a human mind is adequate for forcing a quantum system to collapse into some specific state, then what about the cat's mind? Or the mind of an amoeba? Or a computer? Just what degree of consciousness is needed?)

A less radical suggestion is that consciousness is not actually required—only an interaction between the particle in question and some more complex system. Perhaps simply bumping into run-of-the-mill photons and electrons in the particle's immediate environment is enough to trigger the collapse, nudging the system into a single state. In the case of Schrödinger's cat, for example, the interaction between the radioactive particle and the Geiger counter might be enough to force the system to collapse, giving us a definitely dead cat.

The Many-Worlds Interpretation suggests that every quantum event in which there is a choice of different outcomes actually produces *all* of those outcomes. Why, then, do we see only one result? Because the other outcomes take place in what we might call "parallel universes" that split off from our own whenever a quantum event occurs. In the case of Schrödinger's cat, for example, observers in one universe would see a living cat when they open the box, while those in a parallel universe find a dead cat. What we end up with, over time, is an astronomical number—perhaps an infinite number—of separate universes ("many worlds"). Indeed, anything that *can* happen *does* happen, somewhere in this vast array of universes. There must be a universe where the price of movie tickets keeps going down; there must be another in which I'm dating Julia Roberts. Not surprisingly, the idea of unseen parallel universes has been ridiculed by many scientists; they say it flies in the face of Ockham's Razor, and, worse, defies any sort of experimental test. In recent years, however, a number of prominent theorists have come to embrace the many-worlds view.

Schrödinger's alive-and-dead cat may be purely a thought experiment, but smaller-scale episodes of quantum superposition have shown themselves to be quite real. Early in 2000, for example, physicists at a laboratory in Colorado zapped a beryllium atom with a laser beam, momentarily putting the atom's outermost electron in two quantum states at once. One "copy" of the electron was spinning one way; its twin was spinning the opposite way. The proof that the superposition occurred is that the two copies of the electron interfered with each other, creating a tell-tale signature. The superposition was brief; after about one ten-thousandth of a second, the electron settled into a single quantum state. Nonetheless, such quantum superpositions now appear real—making it more difficult than ever to dismiss the paradox of Schrödinger's cat.

Another bizarre product of quantum mechanics is what physicists call "quantum entanglement"—a peculiar situation in which the state of one particle can appear to influence that of another particle, instantaneously, no matter how far apart they are. This effect had been predicted by theorists in the 1930s, and promptly derided by Einstein as "ghostly action at a distance." In the 1980s, however, such entanglements were shown to be quite real; the effect was first shown by the French physicist Alain Aspect in 1981. Since then, even more dramatic quantum entanglements have been demonstrated. In 1998, for example, researchers at the University of Geneva, in Switzerland, produced a pair of entangled photons. Measuring the energy of one of the photons determined the energy of the other, even though it was 10 kilometers away. The physicists Max Tegmark and John Wheeler, summing up these experiments in a *Scientific American* article commemorating the 100th anniversary of quantum theory, declared: "In short, the experimental verdict is in: the weirdness of the quantum world is real, whether we like it or not."

Today's physicists are still struggling to come to grips with quantum weirdness, as they debate what the theory actually *means*. Or, to be more precise, *some* physicists debate what the theory means; many of them—perhaps most—simply exploit the fact that quantum mechanics *works* and leave the rest to the philosophers. In test after test, the theory consistently provides our best description of the atomic and subatomic world. These scientists have a pragmatic approach: quantum mechanics gives us the probabilities for the different results that a measurement might yield—and that's that. We must simply learn to live with the quantum weirdness that appears to accompany the theory.

Expanding the Quantum Framework

We've seen how quantum theory, born in the first decade of the twentieth century, evolved into quantum mechanics by the end of the 1920s. While experimental physicists eagerly applied the new theory in their investigations of atomic and molecular structure, theorists were working on expanding the scope of the quantum framework. Their greatest challenge was integrating quantum theory with Einstein's special theory of relativity. The result of their efforts—another great step toward unification in physics—was a set of powerful theories known as *quantum field theories*.

One of the first key players in the development of quantum field theories was Paul Dirac (1902–84), a British physicist and mathematical prodigy. He was in his early 20s when he developed a new mathematical framework for quantum mechanics, embracing the formulas of both Schrödinger and Heisenberg. Among the implications of Dirac's equations was the prediction of *antimatter*—a set of fundamental particles with the same mass but the opposite electrical charge of those that make up ordinary matter. The first antiparticle—the antiproton—was discovered in 1954, with exactly the properties predicted by Dirac. A brilliant theoretician but a quiet, introverted man, he eventually became the Lucasian Professor of Mathematics at Cambridge—the position held by Newton three centuries ago and today held by Stephen Hawking. Dirac was awarded the Nobel Prize in 1933. Equally important contributions came from U.S. physicist Julian Schwinger (1918–94) and Sin-Itiro Tomonaga (1906–79) of Japan—as well as another American, the practical-joke-loving, bongo-playing Richard Feynman (1918–88), who elaborated on Dirac's groundbreaking work.

Feynman's innovative pictorial approach to particle physics used simple diagrams rather than complex equations. His pictures, now called "Feynman diagrams," have become vital tools for describing how photons are given off or absorbed by electrons, and have since been applied to a myriad of particle interactions in the quantum world.

Feynman, in spite of once scoring a modest 125 on an IQ test, was indeed an intellectual giant. Still, when *Omni* magazine declared him to be "the smartest man in the world," his mother was reportedly flabbergasted. "If that's the world's smartest man," she said, "God help us."

Schwinger, Tomonaga, and Feynman, who had worked independently, shared the Nobel Prize in 1965 for developing *quantum electrodynamics*, or "QED"—a quantum field theory that describes the interaction between radiation and charged particles. QED was spectacularly successful; it is, in fact, the most precise theory in all of physics. Some of its predictions have been tested in the laboratory to a precision of two parts in 10^{12}—two parts in a thousand billion. If you could measure the Earth's diameter with that kind of precision, you'd be off by no more than a hundredth of a millimeter—much less than the width of a human hair.

Quantum field theories were the great triumph of mid-twentieth-century physics. Indeed, work on these theories is still very much alive, as physicists try to apply them to an ever-broadening array of physical systems. But the greatest value of these theories lies in their success in describing the structure of matter itself. To get a clearer picture of this, we must continue our journey into the atom that we began at the start of this chapter.

A Brief History of the Atom, Part Two

With the work of Rutherford and Bohr in the first two decades of the twentieth century, the atom had given up many of its secrets. Outside, it has a kind of cloud of electrons engaged in a ceaseless quantum dance; inside, protons and neutrons form the tiny nucleus in the atom's center. We saw earlier that opposite charges attract while similar charges repel; this was the starting point in understanding the force known as electromagnetism. That explains why the electrons, carrying a negative charge, stay close to the protons, carrying a positive charge. But why do the protons themselves form such a tight cluster in the nucleus? Why doesn't an atomic nucleus simply fly apart because of the electromagnetic repulsion between its protons? The answer, physicists said, is that a new kind of force must be operating within the nucleus—an attractive force much stronger than electromagnetism. That force, now known as the *strong nuclear force*, is responsible for binding protons and neutrons together and keeping atomic nuclei stable. By the 1930s, however, it became clear that another force was operating within the nucleus. That force, which governs radioactive decay, is now called the *weak nuclear force*.

If you've been tallying the forces so far, you'll see that we have a total of four: the two nuclear forces (strong and weak); the electromagnetic force that we met earlier with the work of Maxwell; and gravity—the most familiar and yet the weakest of all the forces—that we first encountered with Isaac Newton. How many more are there? You may be surprised at the answer. *None.* As far as anyone can tell, these four forces are responsible for every physical process in our universe.

So far, so good: four forces and three kinds of particles— protons, neutrons, and electrons. Plus the photon, the fundamental component of light. A simple picture—but it wouldn't remain simple. Before long, physicists were finding oddball particles that didn't seem to have anything to do with either atoms or light, such as the *muon*, a sort of heavy electron, discovered in the 1930s. As physicists developed more powerful tools for probing the atom in the mid-twentieth century, they found that some of the particles that had seemed "fundamental" were in fact made up of even smaller parts.

You might think that finding these tiny particles requires tiny instruments—but it's just the opposite. In the study of particle physics, the most powerful tool is the *accelerator*, usually a large circular tube in which subatomic particles can be accelerated up to speeds approaching that of light (today's largest accelerators are many kilometers in circumference). Then the particles are smashed into one another, as scientists make a careful record of the resulting micro-fireworks display. The larger the accelerator, the higher the energy levels that can be reached—and the tinier the structures that can be explored.

By the 1960s, particle accelerator experiments had shown that each proton and neutron in an atomic nucleus was made up of three even tinier particles known as *quarks*. The peculiar name was coined in 1964 by U.S. physicist Murray Gell-Mann (b. 1929), who borrowed it from a line in the James Joyce novel, *Finnegans Wake (Three quarks for Muster Mark!)*. The name, he admitted, was a bit of a joke—a reaction against the pretentious scientific language that some of his colleagues were tossing around. There are now thought to be six varieties of quarks in all. Gell-Mann later helped develop the mathematical theory describing how quarks respond to the strong nuclear force. This was another quantum field theory, known by the weighty name of *quantum chromodynamics* or QCD. His contribution earned him the Nobel Prize in 1969.

Soon, more particles were discovered. The *neutrino*, an ultralightweight particle produced in nuclear reactions, was postulated in 1930s and finally detected in the 50s. A particle known as the *tau lepton*—a relative of the electron and muon—was found in 1975. By the 1970s, physicists were able to use the framework provided by quantum field theories to build a coherent picture of how these particles are related to one another and how they respond to the forces of nature. That powerful picture, despite forming the cornerstone of modern particle physics, is saddled with a rather boring name: physicists call it simply the *Standard Model*.

The Standard Model embraces a total of 18 particles. They come in two basic varieties. Twelve of them are *fermions*; these include the electron and the different types of quarks—the particles which make up solid matter. The other six are *bosons*—particles

California Institute of Technology archives

Murray Gell-Mann (left) and Richard Feynman, two pioneers of quantum field theory. Feynman discovered how radiation interacts with charged particles. Gell-Mann discovered how quarks, which make up protons and neutrons, respond to the strong nuclear force.

which "mediate" the interactions between the fermions. (If the Standard Model described global politics, then the fermions would be the presidents and prime ministers, while the bosons would be the diplomatic envoys who shuttle their messages back and forth.) The most familiar boson is the photon; it mediates the electromagnetic force.

This may sound like a painfully large assortment of particles to keep track of: fermions and bosons, quarks and photons. Don't worry about committing them to memory. The important thing is that by the 1970s physicists finally appeared to be isolating the most basic building blocks of matter and energy—the fundamental "bits" that make up the universe. More importantly, they began to understand how the bits relate to one another, through the remarkably successful Standard Model. Further confirmation came as particles predicted by the model were found in the laboratory. The most recent such discovery was a fermion known as the *tau neutrino*, detected at the Fermilab accelerator near Chicago in August, 2000. The previous headline-grabber had been the ultra-heavy *top quark*, snared with much fanfare in 1995, also at Fermilab. So far, more than 20 Nobel Prizes have been awarded to scientists who contributed to the Standard Model.

But it's too soon to celebrate the completion of the Standard Model just yet. One of its most important predicted members, known as the *Higgs boson*, has yet to show itself. This elusive particle, named for the Scottish physicist Peter Higgs (b. 1929), who worked out its properties in the mid-1960s, is supposed to endow other particles with mass. Without the Higgs, in other words, there would be no stars and galaxies; no planets, trees, or people— or, at least, physicists would have to find another explanation for why there *are* such objects. (I met Peter Higgs once at a conference, where at least one appreciative student approached him to say, "Thanks for mass.") The race for the Higgs is now in full swing at particle accelerators like Fermilab in the U.S. and the European Center for Nuclear Research (CERN) in Switzerland. Recent experiments at CERN have offered tantalizing hints of the Higgs—but, as of early 2002, no definitive proof of the particle's existence has been found.

Electroweak Theory: Toward Unification

We could say much more about these various particles and the ingenious experiments designed to snare them. But our focus here is on physicists' efforts to describe what they've found in a succinct, unified picture. The Standard Model was a major advance toward such a concise description. While scientists were developing this model, they were also working out a unified description of two of the four fundamental forces. By the late 1960s, they began to see how electromagnetism and the weak nuclear force could be merged in a single framework. It was aptly named the *electroweak theory*—another step toward the Theory of Everything.

The most important contributions to the new theory were worked out by two U.S. physicists, Sheldon Glashow (b. 1932) and Steven Weinberg (b. 1933), together with Abdus Salam (b. 1926) of Pakistan. (In the late 1940s, Glashow and Weinberg both attended the prestigious Bronx High School of Science, in New York, where they became close friends. The school has the distinction of producing three Nobel Laureate physicists—more than any other school in the world and, indeed, more than many countries.) The physics community, however, was slow to appreciate the merits of electroweak theory, partly because the number "infinity" seemed to keep popping up from its equations and partly because no experiments had yet confirmed any of its predictions. By the early 1970s, however, both hurdles had been cleared. A bright young graduate student from the Netherlands, Gerard 't Hooft (b. 1947), discovered how to get rid of the infinities, and experiments at CERN began to see evidence for the Z particle, one of the bosons associated with the weak nuclear force that was predicted by the theory. Suddenly, electroweak theory was taken very seriously. Glashow, Weinberg, and Salam shared the Nobel Prize in 1979; 't Hooft, together with his colleague and fellow Dutchman, Martinus Veltman (b. 1931), won the Nobel in 1999 for their contributions. When Glashow went to receive his award, he reflected on the progress that theoretical particle physics was making toward unification:

> In 1956, when I began doing theoretical physics, the study of elementary particles was like a patchwork quilt. Electrodynamics, weak interactions, and strong interactions were clearly separate disciplines, separately taught and separately studied. There was no coherent theory

Steven Weinberg helped to unify electromagnetism and the weak nuclear force. He shared the 1979 Nobel Prize with Sheldon Glashow and Abdus Salam.

that described them all. Things have changed. Now we have what has been called a standard theory of elementary physics, in which strong, weak, and electromagnetic interactions all arise from a [single] principle....The theory we now have is an integral work of art: the patchwork quilt has become a tapestry.

Uniting the strong force with the electroweak picture has not been easy. The Standard Model, with its description of quarks—the basic components of atomic nuclei—does embrace the strong interaction, although it is still only poorly understood compared with the electroweak force. However, many physicists believe that at very high energies, the electroweak and strong forces would behave as one. The effort to understand that unification is known as the search for a *grand unified theory*. (The name could be construed as a rather pompous one, especially since such a theory would still leave out gravity. That forced the coining of an even grander name for the yet-to-be-discovered theory that may finally include gravity *and* the electromagnetic and nuclear forces—hence the phrase "Theory of Everything.")

The Theory that Changed the World

Scientists marked the 100th anniversary of quantum theory in 2000, paying tribute to the twentieth century's greatest breakthrough in understanding the physical world. America's leading research journal boasted that "quantum theory is the most precisely tested and most successful theory in the history of science." Indeed it is, and its impact has been felt far beyond the physics lab. Quantum theory has touched almost every branch of science, from chemistry and astronomy to biology and geology. Without quantum theory, atomic structure, chemical reactions, and the forces that hold atoms and molecules together would not be understood. Nor could we explain the radioactivity that heats the earth's core or the nuclear reactions that power the stars. Perhaps the most surprising application came in genetics: without the framework of quantum theory, biologists would never have discovered the structure of the DNA molecules that encode the genetic blueprints of all living things.

Quantum theory has also led to innumerable technological innovations. Without the quantum revolution, there would be no transistors or semiconductors—and therefore no computers, calculators, radios, cell phones, video games, TVs, VCRs, CDs, and DVDs. Indeed, just about every modern technological product, from a microwave oven to a car with a navigation system, has a silicon chip somewhere inside. Nor would we have medical lifesavers such as MRI, CAT, and PET scanners for diagnosing disease and peering inside the body without surgery. Physicist Leon Lederman has estimated that technology based on quantum theory accounts for about one-quarter of the economies of the industrialized world.

Even those bizarre quantum entanglement experiments may prove to have a practical use. The underlying theory may eventually be used to build a "quantum computer." In a conventional computer, a "bit" of formation must be either a "0" or a "1." In a quantum computer, it could be *both* at the same time. Such a computer, physicists speculate, might be able to operate millions of times faster than a conventional computer, solving mathematical problems that today are considered out of reach. (Because of the potential value of such a device in encryption and code-breaking, the U.S. National Security Agency has been one of the biggest financial backers of research into quantum computers.)

True, quantum theory has come at a price. It forced us to abandon the old, secure, mechanistic world-view that had prevailed since the time of Isaac Newton. And the implications of quantum mechanics are just as shocking today as they were in Planck's time. Does quantum theory truly describe our universe? Or is it merely a mathematical "tool" that helps us predict the outcome of certain measurements? And what do we do with the "metaphysical baggage" that seems to accompany it? Such questions, 100 years after the dawn of the quantum, are still open to debate.

Despite its revolutionary nature, quantum theory never captured the popular imagination in the way that relativity did during those early decades of the twentieth century. There are several reasons for this. First of all, while relativity is linked inexorably to the personality and mind of one man—Albert Einstein—quantum theory was the work of many great thinkers working in several countries over a span of three decades. And while relativity clearly applies to realms far from everyday life, quantum theory, one could argue, is even more remote from experience. If nineteenth-century physics began to wrestle science from the amateur's grasp, the first decades of the twentieth century ripped it clean away. "The new science," as historian Jacques Barzun writes, "was no longer within the grasp of the intelligent amateur. Both its concepts and its mathematics required a specially moulded mind, for whom the concepts needed no names but could be read in numerical formulas. This made the scientist still more wonderful but set him as a breed apart." Indeed, the only thing that relativity and quantum theory have in common, to the minds of most people, is that their names are considered roughly synonymous with the word "incomprehensible"—although I hope that in the last two chapters I have helped the reader make at least some sense of both theories.

Quantum theory has taken its place alongside Einstein's general relativity as one of the two great pillars of modern physics. It has also been a powerful aid in the quest for unification—the effort to find a simple explanation for nature's diversity. Though he didn't know how great his contribution would turn out to be, Max Planck himself was a profound believer in nature's ultimate simplicity. "In every important advance," he wrote, "the physicist finds that the fundamental laws are simplified more and more as experimental research advances. He is astonished to notice how sublime order emerges from what appears to be chaos." Over the last century, and particularly over the past few decades, quantum mechanics (along with its modern spin-off, the quantum field theories) has led to a unified view of atomic and sub-atomic physics, including a compelling account of three of nature's four forces.

Quantum theory leaves us with several choices for a celebratory T-shirt, depending on which chapter of the quantum story we wish to honor. For the first half of the twentieth century, the natural choice is the Schrödinger equation, the master formula that governs the behavior of matter at the smallest scales. If we look at the second half of the century, however, we see the great success of quantum field theories and we might choose instead to show off the Standard Model, that collection of fermions and bosons that—as far as we can tell so far—are the fundamental building blocks of our universe.

And yet, the Standard Model has its shortcomings. With 18 supposedly "fundamental" particles, the tapestry that Sheldon Glashow spoke of is still too complicated in the eyes of many theorists. Worse, the model has dozens of arbitrary features, such as the masses of the various particles and the strengths of the forces that govern their behavior. Why, for example, is the electron some 350,000 times lighter than the heaviest quark—with the neutrino lighter still? Today, we can only determine these parameters by measuring them. Physicists would prefer a more elegant theory with an underlying framework that *predicts* those numbers. The Standard Model is "too baroque, too byzantine, to be the full story," says former CERN director Chris Llewellyn Smith. "The Standard Model," concurs Steven Weinberg, "is clearly not the final answer."

As we've seen, the Standard Model is built on quantum theory—a theory that does not involve gravity. Physicists Daniel Greenberger and Anton Zeilinger, evaluating quantum theory in the journal *Physics World*, explain that despite its broad embrace, quantum theory does not reach far enough to explain all of physics:

> When quantum theory does finally break down, as all theories inevitably must, it will be because—in spite of all the strange phenomena it successfully embraces—it will not be sufficiently weird to encompass all natural phenomena. Nature itself is even weirder than quantum theory, and that is sure to be the theory's ultimate undoing.

And so we finish this chapter with the same qualified feeling of success that we reached at the end of the chapter on relativity. We have a spectacularly successful theory, which scientists have learned to apply in a dazzling array of situations—yet in one critical case it fails. Quantum theory does not embrace gravity, for which we must still turn to general relativity. These two great theories—quantum mechanics and general relativity—have long appeared to be mutually exclusive. Like a reluctant bride and groom, they seem to have stopped just before the altar. The next step toward a Theory of Everything must see this marriage through. Today, however, many physicists believe the union is possible. The answer may lie with a strange but beautiful theory that says the universe is made of tiny strings.

Tying Up Loose Ends

String Theory to the Rescue?

From all things the one, and from the one all things.

HERACLITUS

Sooner or later we shall discover the physical principles that govern all natural phenomena.

STEVEN WEINBERG

"Revolution" is not a word that physicists take lightly. They use it only when a new theory brings about a radical change in our picture of the world, challenging the accepted wisdom and offering a broader, more sweeping view of nature. The twentieth century saw two of these revolutions: Einstein's theory of relativity, and quantum mechanics. Today, in the early years of the twenty-first century, some physicists feel that a third revolution is underway. They're excited because of a new physical framework that seems to unite quantum theory and relativity—a new description of nature that may offer a glimpse of the long-sought final theory.

The new approach is called *string theory*. After a turbulent first few decades of life, string theory has firmly entrenched itself as the foremost contender for the Theory of Everything. Two leading string theorists, Edward Witten and David Gross, shared their enthusiasm for the new theory in the pages of the *Wall Street Journal*: "Finally we have the beginning of a new theory that portends to alter our fundamental understanding of physics, to revolutionize our notions of space and time, and perhaps provide the framework for a unified theory of all the forces of nature."

No scientific theory in recent times has had quite as peculiar a history as string theory, which has endured a virtual roller coaster of ups and downs over three decades. It was born in the late 1960s and early 70s, almost by accident, as scientists were poring over the results of high-energy physics experiments in which protons had been smashed together in the world's largest accelerators. Their goal was to develop a clearer picture of the strong nuclear force, the force that binds atomic nuclei together. Mathematically, the theory seemed to describe the most fundamental bits of matter not as point-like particles but as tiny one-dimensional loops of string. The theory generated a brief flurry of activity in the 1970s, but then faded into the background, partly because other approaches to the strong force were showing success and partly because no one knew what to make of string theory's seemingly bizarre predictions. But a handful of physicists stubbornly kept the theory alive—notably John Schwarz of Princeton (he's now at the California Institute of Technology), Michael Green of Queen Mary College of the University of London (he's now at Cambridge), and Joel Scherk of the École Normale Supérieure in Paris. Schwarz, Scherk, and Green played with the new theory's equations and discovered something shocking: string theory seemed to predict the existence of the *graviton*, the particle thought to mediate the force of gravity. A little more mathematical gymnastics showed that Einstein's equations for gravity appear to emerge quite naturally from the string framework.

Suddenly, string theory seemed to be about much more than just the forces within atomic nuclei. It was fundamentally a quantum theory, but after navigating its complex mathematical structures, physicists discovered it actually yields the equations of general relativity. In other words, it appeared to be a true quantum theory of gravity—a potentially far-reaching theory that embraces both quantum theory and gravitation. And so, in the 1980s, string theory again had physicists buzzing.

According to string theory, the universe at its deepest level is made up not of atoms or even quarks, but of these mind-numbingly small loops of string. A typical string size is about 10^{-33} centimeters—so small that it would take a thousand billion billion of them to stretch across an atomic nucleus. If those numbers are hard to digest, think of it this way: each string is as small compared to an atom as an atom is compared to the solar system. But these tiny strings don't sit still; rather, they're constantly vibrating. And, depending on their mode of vibration, they're thought to give rise to the properties seen in the known particles. A string vibrating in one way would appear as a quark;

Human  ~ 100 cm

Magnify 100,000 ×

Today we know that atoms are made of protons, neutrons, and electrons, while protons and neutrons, in turn, are made of quarks. According to string theory, each of these particles is composed of tiny loops of string.

Cell 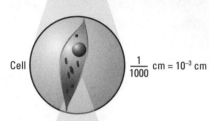 $\frac{1}{1000}$ cm = 10^{-3} cm

Magnify 100,000 ×

Atom 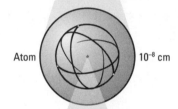 10^{-8} cm

Magnify 10,000 ×

Atomic nucleus  10^{-12} cm

Magnify 1,000 billion billion ×

Strings 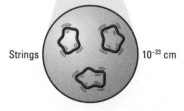 10^{-33} cm

in another, it would appear as an electron. An analogy is often made to the strings on a violin. Depending on how the strings vibrate, any number of musical notes can be produced. String theory seemed to be able to accommodate an infinite number of different particles, none of them more fundamental than the others. (Because of a link between string theory and an idea called "supersymmetry," which we'll hear about shortly, the new theory was also given the label *superstring theory.* To keep things simple, however, I will continue to use the label "string theory" to refer to the original theory as well as its spin-offs.)

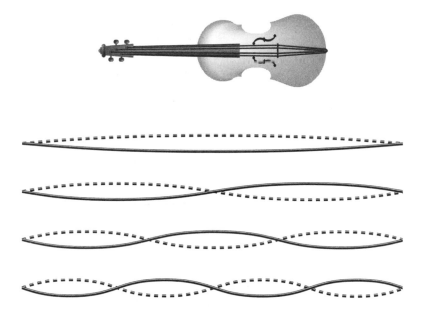

Depending on how they vibrate, the "strings" of string theory are thought to give rise to the properties of the known subatomic particles—just as a violin string can produce any number of musical notes depending on its mode of vibration.

The Theory Formerly Known as Strings

String theory is a modern advance in which scientific discipline?

A Biology   C Chemistry

B Geology   D Physics

By the early 90s, however, string theory was in yet another period of decline, as physicists struggled to make headway with the intricate mathematics the theory demanded. There also appeared to be five different versions of the theory, which led physicists to joke about who was living in those other universes.

The breakthrough came when physicists formulated string theory in 11 dimensions instead of the usual three space dimensions and one time dimension that we see around us. This idea was put forward in the 70s and 80s but didn't gain wide acceptance until the mid-90s. By this time, Edward Witten of the Institute for Advanced Study in Princeton had been at the forefront of string investigations for more than a decade. Witten— colleagues always call him Ed—delivered a series of lectures in which he outlined this brave new 11-dimensional world. He explained that the structures at the heart of the theory need not be one-dimensional strings—they could also be two-dimensional membranes or higher-dimensional "p-branes." ("p" stands for the number of dimensions, so that an ordinary string would be a 1-brane, a two-dimensional membrane would be a 2-brane, and so on.) Witten dubbed the new framework "M-theory," where the "M," he explained, "stands for magic, mystery, or membrane, according to taste." The Internet was soon flooded with papers on M-theory, and the string business was again booming.

According to *Time* magazine, Ed Witten is "widely regarded as the most gifted physicist in the world, and perhaps the most brilliant who has ever lived." In 1990, at the age of 38, he won the Fields Medal—considered the equivalent of the Nobel Prize in mathematics. Witten has always been a modest man, but those around him speak freely of a talent for mathematics that borders on otherworldliness; his students—with great reverence—call him "the Martian." As one colleague put it, "He's brought light where there was darkness. Everything he does is golden." Incredibly, he took a turn in politics and linguistics before zeroing in on theoretical physics for a career—and still managed to make full professor by 28.

Randall Hagadorn

Edward Witten compares string theory to a gold mine of theoretical riches: "We find what looks like a vast subterranean archipelago of gold—everywhere you strike your pick, you find some veins."

I interviewed Ed Witten in 1997. As a thunderstorm raged outside his Princeton office, he told me of his passion for the theory that may one day replace both relativity and quantum theory. He begins with a caveat: like many theorists, he says he doesn't like to use the term Theory of Everything when he talks about string theory. Yet he is clearly enamored with its sweeping potential. "So many remarkable discoveries have been made in string theory, it's hard to believe that the whole thing is just a miraculous accident that has nothing to do with nature," he says in his ethereal, whisper-like voice. "We find what looks like a vast subterranean archipelago of gold—everywhere you strike your pick, you find some veins. You don't really know what's behind it yet, but you know there's something big down there that we've been searching for, for a long time."

Today, with string theory still riding a wave of popularity, Witten's enthusiasm is shared by hundreds of theorists from around the world—and their ranks are growing. "At the moment, we know of only one theory which is viable as a theory of quantum gravity, and

David Gross: String theory may "revolutionize our notions of space and time, and perhaps provide the framework for a unified theory of all the forces of nature."

that's string theory," says Amanda Peet, a young New Zealand-born physicist now at the University of Toronto. "As far as we know, string theory is really the only way of coming up with a fully unified theory that is internally consistent, unifying all the forces and the particles together." The physicist and popular science writer Paul Davies concurs: "I don't know of any other contender for a Theory of Everything that can be taken as seriously as string theory or its M-theory derivative."

Even some of string theory's former critics—including the famous Stephen Hawking—have joined the bandwagon. At one time a vocal critic of the theory (he once described it as "over sold" and "pretty pathetic"), Hawking now devotes much of his research to cosmological models involving string theory and M-theory. Borrowing a phrase from Witten, Hawking says that denying string theory's successes "would be a bit like believing that God put fossils into the rocks in order to mislead Darwin about the evolution of life."

String Theory and Black Holes

If the mass of the star is concentrated in a small enough region, the gravitational field at the surface of the star becomes so strong that even light can no longer escape.

STEPHEN HAWKING

There's so much I don't know about astrophysics. I wish I'd read that book by that wheelchair guy.

HOMER SIMPSON

String theory's next breakthrough came in the mid-1990s, when theorists turned their attention to black holes. Andrew Strominger of the University of California in Santa Barbara (now at Harvard), working with David Morrison of Duke University and Brian Greene of Cornell (now at Columbia), showed that black holes and strings can be described by the same equations. In a sense, black holes can be thought of as fundamental particles in the new framework. But the real coup involved a specific property of black holes known as *entropy*—and it deserves a closer look.

Amanda Peet: "At the moment, we know of only one theory which is viable as a theory of quantum gravity, and that's string theory."

As we saw in the chapter on relativity, black holes form when massive stars collapse at the end of their lives. In a sense, they "disappear" from our universe, leaving behind only their intense gravitational field—a field so strong that not even light can escape. But this view of black holes as a sort of ultimate "cosmic trash compactor" changed in the mid-1970s, with the work of Stephen Hawking.

Hawking holds the distinction of being the only post-Einstein physicist to have become a household name and a cultural icon. Though he is certainly a brilliant physicist, his fame has been bolstered by his popular but frequently unread book, *A Brief History of Time*, and his appearances on TV shows like *Star Trek: The Next Generation* and *The Simpsons*. Hawking suffers from ALS (known as motor-neuron disease in Britain and as Lou Gehrig's disease in the U.S.); he is confined to a wheelchair, and speaks with the help of a voice synthesizer.

Stephen Hawking, in a photo from the 1970s. Once a critic of string theory, Hawking now says "we are on the tracks of the right theory of the universe."

American Institute of Physics

In a groundbreaking 1973 paper, Hawking and Israeli colleague Jacob Beckenstein showed that, because of a quantum-mechanical effect, black holes must actually emit a kind of radiation. If they emit radiation, the scientists reasoned, they must have a temperature. And if they have a temperature, then according to the laws of thermodynamics, they must also have *entropy*—a quantity that measures the degree of disorder in a physical system. (An example from the game of billiards will help: the highly ordered triangular array of balls at the start of a game has very low entropy. After the "break," when the balls are scattered all over the table, the entropy is much higher. A teenager's bedroom, as physicists like to joke, has the greatest entropy of all.) For the first time, Hawking and Beckenstein worked out a formula for the entropy of a black hole.

But a crucial question remained: where did the entropy of a black hole actually come from? Was there any way to understand it at the microscopic level? The answer finally came more than two decades later, in 1996, when Strominger, together with Cumrun Vafa of Harvard, used string theory to count a black hole's quantum states and then used that number to work out the entropy. The formula they found exactly matched the earlier result of Hawking and Beckenstein. The fact that two different methods gave the same result was seen as a clear thumbs-up for string theory.

The black hole result "solves a puzzle that had been around for about 25 years," says Peet, who has worked extensively on the connection between strings and black holes. But the discovery also had another effect. The successful application of string theory to an astrophysical object, she says, forced many of the theory's critics to re-think their objections. "I think that was what finally convinced a lot of people in other areas of physics that string theory actually had something to say about the universe." Even Hawking seems to have had a change of heart. "By now, there are a number of different determinations of the number of quantum states of a black hole," he told a lecture audience in 1996. "They are all very different, but they agree on the answer....In each case, there are a number of objections one could make, but when one has a number of questionable but very different arguments for the same result, one tends to believe it." He later said that the black hole result "greatly increases our confidence that we are on the tracks of the right theory of the universe."

String Theory and the Big Bang

There is a theory which states that if ever anyone discovers exactly what the universe is for and why it is here, it will instantly disappear and be replaced by something even more bizarre and inexplicable.

There is another theory which states that this has already happened.

DOUGLAS ADAMS, *THE RESTAURANT AT THE END OF THE UNIVERSE*

A black hole is a suitable place to test string theory because it is such a peculiar environment. Its gravity is intense—and yet it is a very compact region where quantum effects cannot be ignored. In other words, black holes can only truly be described by a theory that incorporates both general relativity and quantum mechanics—a framework like string theory.

But there's one more realm in which those same conditions can be found: the early universe. As we saw earlier, astronomers believe that our cosmos began with an explosion known as the big bang, some 15 billion years ago. During the earliest moments following that explosion—roughly the first 10^{-43} seconds—the temperature and density of the universe were unimaginably high; indeed, all of the matter that makes up the billions of galaxies we see around us would have been compressed into a region no larger than a proton. At that time, physicists believe, the four forces of nature—the two nuclear forces, electromagnetism, and gravity—were intertwined in one unified force. This unification didn't last very long, however; by the time the universe was one billionth of a second old, it had cooled to a balmy 10^{15} degrees Celsius—still 100,000 times hotter than the core of the sun, mind you—and the four forces went their separate ways.

Cosmologists have been anxious to probe this earliest realm for obvious reasons; the universe we see today, after all, is the direct result of those primordial conditions. At first glance, you might think this urge to look back in time would require a time machine. In a manner of speaking, it does—but not the sort of contraption described in science fiction. Instead, cosmologists make use of the fact that, because of the finite speed of light, we *always* see distant objects as they appeared at some time in the past. In other words, even an ordinary telescope acts as a kind of time machine because of the time required for light to reach us from remote objects. When we look at the sun, for example, we see how it appeared eight minutes ago; we see Alpha Centauri, the nearest bright star to Earth, as it was four years ago; when we peer at the Andromeda Galaxy, we're looking back two million years—and so on. Studies of the microwave background radiation allow us to probe beyond even the range of the largest optical telescopes, while the first gravitational wave detectors, likely to be operational

"It's all string theory to me."

very soon, may tell us even more about those first moments after the big bang.

In theoretical investigations, however, only a theory that combines relativity and quantum theory can hope to explain the properties of this ultra-compact infant cosmos. That's why many cosmologists are beginning to use ideas based on string theory to explore the physics of the big bang. We'll hear about some of these bold string-based models of the cosmos shortly. But first we need to take a closer look at the big bang model of the early universe.

For decades, astronomers had noticed that the universe appeared to have the same properties in every direction; no matter where they aimed their telescopes, they saw roughly the same density of galaxies. Even the microwave background radiation is remarkably homogeneous, displaying only tiny ripples over an otherwise strikingly smooth backdrop. This "smoothness problem" has long puzzled physicists, because the finite speed of light should have made it impossible for widely separated regions of the universe to have had any "contact" with each other. To use a rough analogy: suppose you're a schoolteacher and you find that two students sitting next to each other hand in the exact same answers on a test. Obviously, you conclude, one student copied from the other. But suppose *all* the students have the same answers. Even if each student copied from his or her neighbor, it would have taken some time for the information to percolate across the entire classroom. If the room is large and the time allotted for the test is short, the similarity of the test results becomes quite puzzling. The problem was finally solved in 1980 by cosmologist Alan Guth of Stanford University (he's now at the Massachusetts Institute of Technology). His solution involved a modification of the traditional big bang picture. The new theory was called *inflation.*

According to the inflation model, the universe underwent a tremendous, exponential growth spurt in the first moments following the big bang—roughly from 10^{-35} to 10^{-33} seconds following the initial explosion. During that time, the universe ballooned from about 10 billion billion times smaller than a proton to about the size of a grapefruit—a leap in size amounting to a staggering factor of about 10^{50}. In the inflation picture, the universe was originally small enough for the smoothness "information" to spread easily from every region to every other region. (In the classroom example, imagine how much easier it would be for the students to cheat if they were crammed into a small classroom, rather than spread out in the gymnasium.) Inflation wasn't perfect, however, and various modifications were proposed over the next few years. Still, by the 1990s, most cosmologists had embraced the basic idea of the inflation model, and virtually every measurement and observation agrees with its predictions.

In the spring of 2001, however, Princeton physicist Paul Steinhardt and his colleagues proposed an alternative to the inflationary model known as the "ekpyrotic universe" (the name comes from a Greek word meaning "cosmic fire"). This model is based on the extension of string theory that we heard about earlier, in which matter is made up of multi-dimensional membranes rather than one-dimensional strings. Though it sounds a bit surreal, the ekpyrotic model describes our universe, with its three familiar spacial dimensions, as a giant membrane (a "3-brane") floating alongside other "parallel" branes in a multi-dimensional cosmos. In this picture, the big bang is re-cast in an entirely new light. Steinhardt's model describes the big bang as a collision between two adjacent membranes—a kind of cosmic "splat" produced by two giant sheets slamming together.

How does the ekpyrotic model account for the smoothness of the universe? Because all of the membranes are slow-moving, Steinhardt says, there's plenty of time for information to travel across vast distances (in our classroom example, this would be like extending the time allotted for the test). It also appears to have an advantage over the inflationary model: in the inflation scenario, the actual moment of the big bang is home to a "singularity"—a troubling state in which the temperature and density are infinite and defy description by any conventional theory. In the ekpyrotic model, however, the universe is never confined to a single point in space-time, and there's no singularity to reckon with.

Steinhardt and his colleagues have recently developed a model that takes the ekpyrotic scenario a step further: not only is there no singularity, but there is no "beginning" to the universe. Instead, he proposes a "cyclic universe" in which there are a possibly endless series of expansions and contractions as these branes bounce back and forth in a multi-dimensional cosmos. In this model, the event that we think of as the big bang would simply be a transition between two phases in what may be an eternal universe. "The central issue in all this," says Steinhardt, "is the question of what actually happens at the big bang. Is it that, as we go backwards through time, that time actually comes to an end? Or is time continuous, and it's just a transition to a pre-existing phase? Ultimately, I think string theory can answer this." Steinhardt says that experiments may one day allow scientists to distinguish between the predictions of the two models—most likely by using gravitational wave detectors to study the primordial "gravitational imprint" of the big bang, or by examining the cosmic microwave background for the effects of the gravitational waves that washed across the early universe. Such measurements, however, may still be many years away.

So far, the ekpyrotic model has received a mixed reception; while a number of cosmologists welcome an alternative to inflation, they caution that inflation is far more established. Andrei Linde, who made important contributions to the original inflation model, says the ekpyrotic scenario is unnecessarily complicated, like the epicycle-ridden medieval cosmos. Still, Steinhardt's proposal may herald a new phase in theoretical cosmology: for the first time, scientists studying the universe-at-large are taking a keen interest in string theory and are making a concerted effort to apply "string ideas" to the cosmos itself. This sort of dialogue between the cosmologists and the string theorists, says Amanda Peet, is quickly becoming routine. Even within her own institute, she says, "we have a bunch of people in the cosmology and gravity program, right from the experimental end all the way to string theory, who are talking to each other." Glenn Starkman, a particle physicist and cosmologist at Case Western Reserve University in Cleveland, agrees that string theory has become so prevalent that it's now impossible to ignore. "When I work on things having to do with fundamental theory," he says, "I try to make sure they're either string-inspired or at least string-compatible."

Putting String Theory to the Test

String theorists aren't uncorking the champagne just yet, however. The biggest problem is that the theory describes how particles and forces behave at enormously high energies. Even the largest particle accelerators are about a billion billion times too weak to probe string effects by any direct experiment. Facilities like Fermilab and CERN are fine for producing particles like quarks and muons, but hopelessly inadequate for detecting strings. (To investigate strings directly, in fact, would require an accelerator roughly the size of the solar system. Obviously such a facility is unlikely to receive funding in the foreseeable future.) Even the breakthrough with black hole entropy was hardly an "experiment" (the nearest known black hole, after all, is more than a thousand light-years away); it was simply an agreement between two highly theoretical studies. Without a direct test for string theory, skeptics argue, it should be classified as mathematics rather than physics—at least until the theory makes predictions that can be tested in the laboratory and not just on the blackboard. Even those who cautiously endorse string theory are careful not to pin their hopes on an idea that has yet to confront experiment.

"I don't think it's physics yet," says Glenn Starkman. "It's beautiful mathematics, which has a chance of becoming physics." So far, he says, "no one has figured out how to make any predictions that can be tested." Starkman concedes, though, that string theory is head-and-shoulders above any of its rivals. Competing theories do exist: there's "loop quantum gravity," for example, which postulates that space itself, rather than matter and energy, is composed of tiny loops; and "twistor theory," an offshoot of general relativity in which rays of light, rather than points in space, are seen as fundamental. Both are far from being fully developed theories, and a detailed discussion of them is beyond the scope of this book. String theory, for the moment at least, is the one in the limelight and the one drawing the brightest graduate students. While string theory has not yet earned the label Theory of Everything, Starkman says, "It's the only game in town."

One of string theory's most outspoken critics has been Sheldon Glashow, a Nobel Laureate physicist at Harvard. (We heard about his role in developing the unified electroweak theory in the previous chapter.) He says that string theory says nothing—at least so far—about the real world. "I'm very happy that so many of my colleagues are working on string theories, because it really keeps them effectively out of my hair," he said in a BBC Radio interview in 1987. He added, presumably tongue-in-cheek, that he does everything in his power "to keep this contagious disease ...out of Harvard," but so far he has "not been very successful." When I interviewed Glashow ten years later, he was still skeptical. The theory lives at enormous energies, he explained, making it nearly impossible to verify. "String theory, if it's to be acceptable, must confront experiment; it must reproduce what we already know," Glashow says. Perhaps it gives you gravity, he acknowledges, but that's not enough. It has to predict the strong, weak, and electromagnetic interactions as well. "So far, string theory doesn't do this. It lives in the sky, so to speak."

String theorists are hopeful, however, that their theory can be brought down to earth after all. The strings themselves may be out of reach, but less direct tests may be quite possible. One of their greatest hopes is pinned to an idea known as *supersymmetry*.

Supersymmetry and Hidden Dimensions

Supersymmetry refers to a suspected but very subtle symmetry in the laws of nature that appears to link the two kinds of particles that we met in the previous chapter—the fermions that compose matter and the bosons that mediate the forces between them. The idea, which dates back to the birth of string theory three decades ago, says that every fermion has a boson partner and vice versa. More specifically, supersymmetry predicts a new set of fundamental particles—one *superparticle* corresponding to each of the known particles. These superparticles, which would be more massive than their ordinary partner particles, have never been observed. However, the next generation of giant particle accelerators may be powerful enough to detect them. Because supersymmetry has now been incorporated into string theory, its proponents now eagerly await these new accelerator experiments.

While strings, as Glashow said, "live at enormous energies," these superparticles can be investigated at much lower and more readily accessible energies. The best bet for observing supersymmetry may lie with an accelerator known as the Large Hadron Collider (LHC), now under construction at CERN, near Geneva. (A hadron is any particle that responds to the strong nuclear force.) The $4-billion detector will be about seven times more powerful than today's largest accelerator, the Tevatron facility at Fermilab. (Funding for a similar accelerator in Texas, known as the Superconducting Supercollider, was cut by the U.S. Congress in 1993.) The LHC will smash a billion or so protons into one another each second, possibly creating more massive and as yet unknown particles. This will be the closest that scientists have come to recreating the conditions of the big bang, and will allow physicists to probe the structure of matter down to the scale of 10^{-21} centimeters. The LHC should have enough clout to detect superparticles—if they in fact exist. "If they turn on the LHC and discover the supersymmetric partner of the electron, that would be terribly exciting," says Amanda Peet. If that happens, "a lot of people would think that string theory was likely to be the right theory of the world."

Supersymmetry may not be the only aspect of string theory that can be investigated in the laboratory. As we saw earlier, string theory seems to work best when it's formulated in 11 dimensions (though some formulations only require 10). Don't be alarmed if you can't picture such a plethora of dimensions; no one else can either. The extra dimensions are required mathematically in order for the equations to be self-consistent. If these additional dimensions exist in the real world, physicists say, they elude detection because they're curled up in a scale too small to be seen. As an example, think of a garden hose (see diagram). From a distance, it looks one dimensional, like a simple line; any point along its length looks like just that—a point. But as we move closer, we see the hose actually has another dimension. What we thought was a point turns out to be a circle. Notice, however, how different these two dimensions are from each other. An ant crawling along the length of the hose can travel a great distance; an ant crawling along the "circular" direction can go only a short distance before returning to where it started from. Physicists speculate that the extra dimensions of string theory may follow a similar pattern: while the three space dimensions and one time dimension are large (possibly infinite), the additional seven dimensions likely fold back on themselves over very short distances—probably over a scale smaller than the size of a proton.

String theorists were not the first to suggest that we live in a universe with more than four dimensions. Back in the 1920s, the Polish physicist Theodor Kaluza and the Swedish physicist Oskar Klein proposed a five-dimensional framework in an attempt to

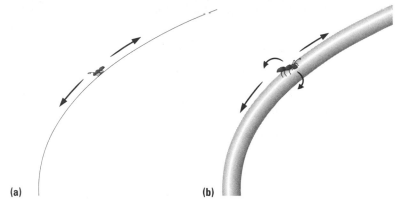

(a)   (b)

Viewed from a distance, a garden hose appears one dimensional, but an ant crawling on its surface could tell that it has another dimension—the hose's circular cross-section. According to string theory, our universe may contain "hidden dimensions" that are curled on such a small scale that they elude detection.

unify Maxwell's electromagnetism with Einstein's general relativity. Although a few physicists, including Einstein, were intrigued by the mathematical elegance of the theory, most were simply baffled by it; they saw no reason to suspect it described anything real. Later, when the strong and weak nuclear forces were discovered, the Kaluza-Klein theory seemed inadequate, so the idea of extra dimensions was largely abandoned—at least until supersymmetry and string theory began to take shape a half-century later.

The question for string theorists is whether these hidden dimensions can be made to reveal themselves without having to actually probe the impossibly small scales of the strings themselves. The key may lie in the behavior of the most familiar of the four forces—gravity. In Newton's theory, the force of gravity falls off with distance according to the inverse-square law. Newton showed that this law described the motion of the moon as well as the planets and comets in our solar system; later astronomers saw that it applied just as well to multiple-star systems light-years from Earth and even to the motion of galaxies and galaxy clusters millions of light-years away. Newton also showed that the same law governed the way falling objects respond to gravity here on Earth—hence the label "universal gravitation."

In the late eighteenth century, British physicist Henry Cavendish probed smaller scales by carefully measuring the attraction between pairs of lead balls about 20 centimeters apart; the result again supported the inverse-square law. But how does gravity behave over still smaller distances? What happens if the masses are only a centimeter apart? Or a millimeter? Exploring gravity at these scales is fiendishly difficult, but the results of such experiments may prove immensely valuable. If string theorists are right about hidden dimensions, gravity may actually deviate from the inverse-square law over such tiny distances.

A number of experiments designed to test the behavior of gravity at the millimeter and sub-millimeter scale are now underway in U.S. and European labs. "If there were a fifth dimension at the scale of a millimeter, you would expect the inverse-square law to become an inverse-cube law when you got within the size of that fifth dimension," says string theory pioneer John Schwarz. If an inverse-cube law prevailed, doubling the distance would reduce the strength of gravity by a factor of eight—two cubed—rather than the usual factor of four. "As yet, there's no evidence for anything of that sort—but the experiments can still be pushed farther. It's a very fundamental question to be explored experimentally."

In studying these extra dimensions, physicists and cosmologists have focused in particular on the idea of "branes"—the extension of string theory into higher dimensions that we heard about earlier. In some versions of the brane picture, the extra dimensions may turn out to be very large rather than very small. As we saw with Paul Steinhardt's ekpyrotic model, some scientists have even proposed that one or more hidden dimensions actually surround the universe we live in and may be infinite in size. The brane picture, physicists speculate, may explain why gravity is so weak. While the other three forces are "trapped" in our familiar four-dimensional world, gravity might "leak out" into one of the hidden dimensions. It could also explain the

Lisa Randall and Raman Sundrum. Their string-inspired cosmological models might explain why unifying gravity with the other forces of nature requires such high energy levels.

problem of "dark matter"—the puzzle of why much of the universe seems to be made of something other than the familiar luminous matter that forms stars and galaxies. The missing matter, they muse, could be trapped on one of these other branes, with its gravity—but not its light—slipping through to our universe.

An intriguing variation involves just one extra, infinite dimension that's curved or warped rather than flat. This idea, put forward by physicists Lisa Randall of Harvard and Raman Sundrum of Stanford (now at Johns Hopkins University), begins with a premise similar to Steinhardt's—that our universe "is confined to a kind of lower-dimensional island within higher-dimensional space," as Sundrum explains it.

If the Randall-Sundrum picture is correct, it might explain what physicists call the "heirarchy problem"—the puzzle of why the electromagnetic and nuclear forces seem to merge at an energy level so much lower than that required for the more complete unification, embracing gravity. The theory also predicts certain exotic new particles that may be detected at the LHC before the end of the decade. The theory, Randall says, "has opened up an entirely new branch of 'beyond the Standard Model' physics."

All of these ideas, of course, are highly speculative—but they are far from being the most speculative ideas in theoretical physics. All of them offer at least the potential for experimental support. And all of them are tied together by their link to string theory—a theory which, despite its seemingly peculiar view of the world, is for now the most promising path toward a Theory of Everything.

The Beauty in the Equations

Mathematics possesses not only truth, but supreme beauty—a beauty cold and austere, like that of a sculpture ...sublimely pure, and capable of a stern perfection such as only the greatest art can show.
BERTRAND RUSSELL

Any physicist will tell you that the ultimate hurdle for a new theory is experimental testing. A theory whose predictions are supported by experiment is seen as either "likely to be true" or, at the very least "possibly true"; a theory refuted by experiment is worthless and soon discarded. Of course, there are always those who cling to the old paradigm, and the rise and fall of theories is not necessarily something that happens quickly (as Thomas Kuhn and a host of other philosophers and scientists have pointed out). Still, theories that contradict experiment eventually wind up in the historical dustbin. As philosopher Bas van Fraassen has put it, scientific theories "are born into a life of fierce competition, a jungle red in tooth and claw"—and only those that agree with experiment survive.

But throughout our story, and particularly in this chapter, we've seen another guiding principle at work—the idea of mathematical "beauty" or "elegance." It's hard to describe exactly what physicists and mathematicians mean when they use these terms, just as it's hard for artists and musicians to explain why they're drawn to certain designs or melodies. And yet this mathematical beauty is there, just as it is for the artist, and it is undeniably a factor in the physicist's search for simplifying, unifying descriptions of nature.

We saw this principle at work in the ideas of Copernicus and Kepler, as they struggled to understand the nature of the solar system. When describing his sun-centered model—far more elegant, in his eyes, than the Ptolemaic model it would replace—Copernicus spoke of the "marvellous symmetry of the universe and a firm, harmonious connection between the motion and the size of the spheres...." When Kepler discovered a simple mathematical law describing planetary orbits, he said that his model of the heavens left him "filled with unbelievable delight at its beauty." Hans Christian Oersted, who discovered the link between electricity and magnetism, wrote a book called *The Soul in Nature*, about the harmony between the beauty in science and that found in art and music. In the twentieth century, we saw how Einstein, always passionate about geometry, was just as enamored with the equations of general relativity, calling them "beautiful beyond comparison." Werner Heisenberg, one of the founders of quantum mechanics, called his equations "strangely beautiful" and spoke of the "wealth of mathematical structures" the theory contained. Paul Dirac mused that "it is more important to have beauty in one's equations than to have them fit experiment."

Today we hear similar sentiments in support of string theory. We heard Ed Witten comparing string theory to a gold mine of theoretical riches; Cumrun Vafa, who used string theory to measure black hole entropy, says the theory "is the most beautiful and

John Schwarz: "The fact that mathematics is the right language for doing this ...is a deep truth that I don't think anyone really understands."

consistent structure that we have." And Columbia string theorist Brian Greene likely had this notion of beauty in mind when he chose the title of his recent book, *Elegant Universe*. Of course, there are dangers in letting aesthetics drive one's research program. Aristotle, Ptolemy, and even Copernicus were so certain that planetary orbits were circular that they did not look into alternatives. Kepler eventually found the true shape—the ellipse—but, as we saw, his fascination with the five "regular solids" of geometry briefly led him astray.

Even Einstein, late in life, may have put too much emphasis on mathematical beauty; he referred to mathematics as "the only reliable source of truth" as he struggled in vain to formulate a unified field theory. Einstein biographer Albrecht Fölsing cautions that "just as the theoretical physicist is helpless without mathematics, so mathematical speculation unrelated to reality remains void." The aging Einstein, he says, displayed a "reckless overestimation of the possibility of understanding nature through mathematics alone." Kepler eventually realized his error and recovered; Einstein did not. Nonetheless, at the risk of greatly oversimplifying the case, mathematical elegance provides a fruitful guide (though of course not the only guide) in developing physical theories. No one can deny that beautiful theories have an established track record of outperforming "ugly" theories (though again we must remember that we have no unambiguous definitions of these terms).

Of course, the question of *why* mathematics is such a useful tool for the physicist is another matter. Most scientists would simply agree with the Hungarian mathematician and physicist Eugene Wigner, who said that the appropriateness of mathematics in describing nature "is a wonderful gift which we neither understand nor deserve." John Schwarz, echoing Galileo's famous remark on mathematics being the language of nature, says:

> It seems to be a profound truth that important theoretical advances in physics, dealing with fundamental issues, tend to have elegant answers. The fact that mathematics is the right language for doing this stuff is already a deep truth that I don't think anyone really understands; they just know that it's so. I find it profound.

In his book *Dreams of a Final Theory*, Steven Weinberg puts it this way:

> Time and again physicists have been guided by their sense of beauty not only in developing new theories but even in judging the validity of physical theories once they are developed. It seems we are learning how to anticipate the beauty of nature at its most fundamental level. Nothing could be more encouraging that we are actually moving toward the discovery of nature's final laws.

It's worth adding that just as beauty can suggest a promising path, lack of beauty can alert a physicist that something is amiss. In the previous chapter, we saw how the Standard Model accounts for the diversity of particles seen in nature—yet physicists balk at its large number of arbitrary parameters. In short, they say it's too inelegant, too ugly, to be the final answer.

The Eye of the Beholder

All of this talk of mathematical beauty and elegance may at first seem paradoxical. A new theory might be touted as giving a simplified view of physics—but with all the esoteric mathematics involved, it might not *look* simple at all. We must remember, though, that mathematical beauty, like the beauty seen in art, is in the eye of the beholder—and the trained beholder at that. A particle physicist who looks at a set of equations may well see a structure of great beauty. For the rest of us, it may be difficult to see *any* beauty in *any* of the equations. When I asked physicist Leon Lederman about this, he drew a comparison between physics and art. Just as a Cubist painting would have baffled Titian or Rembrandt, so the equations of string theory would have held little appeal for Newton or Descartes:

> Beauty is sometimes in the eye of the trained beholder. I think that it takes practice and it takes some experience. And this is true even of the beauty you see in art. You can imagine a seventeenth century painter being confronted with modern, abstract art. He would not see the beauty that people who have grown up with art through its various phases and changes would see....It's the same with music. There's modern music, atonal music of various kinds, which is greatly appreciated by people who are trained in it—but which would have been bizarre to composers, say, at the time of Beethoven or Mozart. So it takes some training. And especially it takes some mathematics, because often when we look at the beauty, it's because these theories have to be written down in mathematical form. And the mathematics itself—to people with some training—can be expressed as being beautiful and simple in many ways.

As Lederman cautions, what appears ugly or intimidating today may seem second-nature, or indeed quite beautiful, to later generations. This happens so often in the world of art that we are rarely surprised to hear that some particular work, acknowledged today as a masterpiece, was lamented or even ridiculed when it was first created. (The example of the Eiffel Tower comes to mind. When it was built in the 1880s, critics denounced it as "monstrous," "a barbarous mass," a "stupefying folly," and an

"Oh, if only it were so simple."

"odious column of bolted metal"; today, of course, it has become a symbol of Paris and an architectural icon.)

Yet this "training" that Lederman spoke of, whether in art or science, does not come without great effort—and a great deal of mathematics may be required in order to truly appreciate a modern physical theory. Anyone who's taken a university physics course knows this very plainly; to master the physics, you have to master the math, and the farther one advances, the more math one needs to learn.

This pattern is reflected throughout the history of physics. When Isaac Newton was studying motion, he realized he needed a new set of mathematical tools—so he invented calculus. Einstein got off slightly easier: when he was formulating general relativity, he found that Bernhard Riemann had developed a geometry of curved spaces just in time to serve as the framework of his new

Critics once denounced the cast-iron bulk of the Eiffel Tower as an eyesore. But as the use of iron became more widespread—and thus more familiar—people began to see the tower's novel design as beautiful. New theories in physics, often highly mathematical, face a similar challenge.

(Author photo)

theory. Today, the string theorists are in a position that combines the predicaments of both Newton and Einstein. They're busily learning esoteric branches of mathematics—you'll hear them talk about topology, homotopy, cohomology, Lie algebras, and the like—and when such methods aren't powerful enough, they're forced develop new techniques, boldly taking their equations where no math has gone before.

Even the brightest of physicists can find the details of string theory difficult. "I must confess," says John Schwarz, "I've found myself struggling to learn areas of mathematics I never conceived I would need to know." Freeman Dyson, a brilliant physicist who made important contributions to quantum electrodynamics, recalls a lecture Ed Witten gave at Princeton in the 1980s. After the 90-minute talk, the audience was left in stunned silence. "There were no questions," Dyson recalls. "Not one of us was brave enough to stand up and reveal the depths of our ignorance." Leon Lederman, who won a Nobel Prize for his work in particle physics, had a similar experience when Witten spoke at Fermilab. "I rushed over to my lab to explain what I had learned to my colleagues," Lederman says, "but by the time I got there I had lost most of it."

Will it ever get easier? Perhaps—once string theory is more fully developed and the mathematical techniques are more clearly understood. According to Witten, the struggle that string theorists are facing parallels the mathematical struggles that challenged

Einstein and the other great innovators in the history of physics. General relativity, for example, at first "dazzled people with its complexity," he says. "But then they came to understand it better and to appreciate the beauty." The same thing, Witten says, happened with Maxwell's theory in the previous century: "Maxwell's equations of electromagnetism, which we now teach to freshmen, were—in the late nineteenth century—considered so difficult and complex that North Americans who wanted to learn them had to travel to Europe." (Again we see a parallel in the arts; the Eiffel Tower, for example, earned the label "beautiful" only when the utility of its novel building material—cast iron—was realized, and its use became more widespread and thus more familiar.)

The most important point here—one which all string theorists acknowledge—is that string theory, along with its M-theory and p-brane derivatives, is still "under construction." Once it is better understood, theorists hope its inner beauty will be revealed for all. Eventually, Witten says, it will be taught to undergraduates, just as relativity and quantum mechanics are today. As physicist and author John Barrow puts it:

> Quite often when you do mathematics and physics, your first way of doing something is unnecessarily complicated and difficult. It makes use of the tools that you're familiar with, rather than the ones that are best suited for this unfamiliar job. What often happens is that after you have constructed the new building, as it were—with great effort and in a rather ungainly way—then people will go back and re-build it, using better mathematical tools, and there will be a much simpler way to do it. You'll look at it from the right direction and think, 'Well, if only we had looked at it in that way from the start, we could have got there so much more easily.' So the fact that things look to be so complicated might just be a consequence of this being a work in progress.

Barrow says he's confident that this process of simplification will eventually happen with the Theory of Everything—whether it turns out to be a form of string theory or something quite different yet to be discovered. "If we do find a good description of a 'theory of everything,' there will be simpler ways to explain it," he says. In the end "there will be ways to make it reasonably accessible in its key ideas to ordinary members of the public." Ivan Tolstoy, author of a popular biography of Maxwell, agrees. Once a theory is fully developed, the average person will be able to "see through" the mathematics to the underlying simplicity it is founded on:

> The physicist's tools—his mathematical language, his experimental know-how—bear the same relationship to his thought as the brushwork of the artist bears to the whole painting. They are important. Yet the lover of art can extract the essence of a Rembrandt without understanding or even caring much about technique. The message is accessible to millions because it transcends its craft.

The Unfinished Revolution

Because string theory remains unfinished, it's not surprising that it still has its critics. String theory is "a long shot" that has "taken over at the expense of all other areas," says renowned mathematical physicist Roger Penrose of Oxford. "Some of the mathematical notions which people associate with string theory are very appealing. But just because they are appealing doesn't mean they are right." The most common objection, as we've seen, is that string theory has yet to directly confront experiment. Dutch physicist Gerard 't Hooft, who won a Nobel Prize for his contributions in unifying the electromagnetic and nuclear forces, cautions that string theory has been "deafeningly silent" in explaining specific physical or cosmological parameters. So far, "string theory has been unable to predict anything," he says, adding that "string theorists [should be] modest about their 'breakthroughs.'"

String theorists, meanwhile, are quick to admit that their theory, even in its latest M-theory incarnation, is far from complete. They've discovered dozens of compelling mathematical structures, they say, but have yet to unearth the fundamental principles that underlie the equations. "There is still a gaping hole at the center of the M-theory jigsaw where we don't know what is going on," says Stephen Hawking. John Schwarz says that "we are still quite far from a complete understanding of this marvellous mathematical edifice."

Much of the difficulty may stem from the fact that string theory, as we've seen, was discovered by accident. Back in the 1980s, Ed Witten often described string theory as an example of twenty-first-century physics that had accidentally fallen into the hands of twentieth-century physicists. Or, as Jeff Harvey of the University of Chicago puts it, the discovery of string theory is "like a primitive society finding some advanced tool left behind by some other culture. You push this button and it does one thing, and you push another button and it does something else. By pushing different combinations of buttons, you start to understand that it's very powerful and can do all sorts of interesting things"— and yet an understanding of *why* it works, and how far it can reach, remains elusive. String theory may illuminate the path, but we have not yet reached our destination.

This is the point in our story where I would unveil the "ultimate T-shirt"—the set of equations describing how the universe is put together, in a concise, elegant formulation—if such equations were known. So far, they are not. The best we can do is to display *two*

great descriptions of the cosmos: general relativity, which describes the macroscopic world from terrestrial mechanics to the motion of planets, stars, and galaxies; and quantum mechanics, which describes the microscopic world of quarks and electrons and the other residents of the subatomic zoo. String theory, including its more elaborate M-theory spin-off, seems to offer the best hope for unifying these two theories—of tying them together, so to speak. But string theory, for now, remains incomplete. Proponents are drawn by its promise of a quantum description of gravity and by its mathematical "beauty"; critics say it has yet to make any concrete, testable predictions. It is not yet the final theory.

In the meantime, theoretical physicists are doing what they do best: speculating on how the universe *might be*, based on what they know about how the universe *is*. They have a powerful and ever-expanding mathematical tool kit to aid their investigations, and a keen sense of aesthetics to help them select the most promising pathways. Many of their ideas, perhaps most, will turn out to be wrong—in some cases, laughably, embarrassingly wrong. But there is always the possibility that some of these newly developed ideas will bear fruit. Some might even point the way toward the Theory of Everything—and that possibility warrants occasional forays into the realm of educated speculation. As Albert Einstein said: "The theorist should not be carped at as 'fanciful'; on the contrary, he should be granted the right to give free rein to his fancy, for there is no other way to the goal. His is no idle daydreaming, but a search for the logically simplest possibilities and their consequences."

What Does It All Mean?

Science, God, and the Limits of Understanding

In reality we know nothing, for truth is in the depths.

DEMOCRITUS

The eternal mystery of the world is its comprehensibility.

ALBERT EINSTEIN

Imagine the year is 2050, and a brilliant young physicist—perhaps your granddaughter or grandson—has just put the finishing touches on the Theory of Everything. What happens next? A parade on Fifth Avenue—or just a few drinks at the campus pub? The particle physicists, of course, would be thrilled. The new theory would finally explain all those seemingly arbitrary parameters in the Standard Model. The charge on an electron, the mass of a muon, the strengths of the four forces—all of these things, and many more, will suddenly make sense. The new theory will *require* those parameters to have the specific values observed in nature. The complexity of particle physics—all of those quarks and leptons and bosons—will give way to a more concise, elegant picture founded on the new theory.

A unified theory of physics will also have a profound impact on cosmology. It will illuminate that epoch—those earliest moments after the big bang—when the temperature was so high that the four forces would have been unified as a single entity. Perhaps the new theory will explain what caused the big bang itself. The Theory of Everything will also explain the origin of Einstein's "cosmological constant"—the most likely candidate for the mysterious anti-gravity force that we heard about earlier. Because this force (together with gravity) directs the evolution of the cosmos, a true unified theory will also help us predict the ultimate fate of the universe.

So let's say that a great new theory has done all that. It's made the particle physicists happy and it's made the cosmologists happy. Newspaper headlines proclaim a Theory of Everything is at hand; its discoverer is hailed as the next Newton or a second Einstein. Would it be the end of science? Or at least the end of physics? To answer that question, we must take a closer look at the kinds of explanations that science offers.

Reductionism vs. Holism

Science cannot tell us a word about why music delights us, or why and how an old song can move us to tears.

ERWIN SCHRÖDINGER

For centuries, much of science has been tied up in a philosophy called *reductionism*. The idea is that if you understand the parts, you understand the whole. A biologist, for example, can learn a great deal about living organisms by studying their genes. To understand the DNA that makes up an individual gene, we consult the biochemist, who looks at its molecular structure. To understand the carbon, nitrogen, oxygen, and other elements that make up those molecules, we ask the chemist—and so on, down the line, all the way down to the particle physicist. Now suppose that the particle physicists have found a Theory of Everything. Will that explain what the chemists and biologists see in their laboratories? A die-hard reductionist would say that it should—after all, the Theory of Everything would describe the ultimate "parts" that make up the whole.

Reductionist science, in fact, is one of the great success stories of the last 400 years. Its track record is stunning. It explains why copper and gold conduct heat and electricity better than wood and rubber; it explains the properties of liquids and gases in terms of their component molecules; it explains how water freezes and boils, why iron rusts, why candles burn, and why glue is sticky—and much, much more.

Yet there are many systems, even in physics, where the reductionist approach fails—systems in which the whole is more than the sum of its parts. The example mentioned most often is weather forecasting. In principle, we already know how wind and rain and snow work. Yet if we try to predict development of a tropical storm, for example, we run into trouble. The system is simply too complex to lend itself to precise predictions. (Of course, weather forecasting *has* improved over the years, thanks to innovations like satellite imaging—but a deeper knowledge of the atoms that make up a hurricane will not take us any further.) In the life sciences, we see even more examples. The behavior of animals, the evolution of ecosystems, the spread of disease— studying any of these requires a knowledge of the whole system as much as an understanding of the component parts.

In human affairs, a Theory of Everything will accomplish even less. It will do little to explain love, hatred, humor, music, or war— simply because physics has relatively little to say about such matters in the first place. It all boils down to the "level of description" we're after. A Beethoven symphony *could* be described in terms of sound vibrations of various frequencies and intensities, and *Hamlet* might be interpreted as a pattern of black marks on a white page—but one would learn very little from such a description.

In science, most systems lie somewhere in between. A description at the subatomic level is crucial for some systems—virtually anything involving chemistry, for example—but irrelevant for others. As physicist and author Paul Davies has put it, a Theory of Everything "would not tell you how life originated, it wouldn't tell us about the nature of consciousness, and it wouldn't tell you how I'm going to vote in the next election." Even within physics, there is a world of difference between *discovering the laws* that govern a physical system and *applying those laws* to make specific predictions. A comparison is often made to the game of chess: learning the rules takes just a few hours; becoming a grand master can take a lifetime.

Whether a Theory of Everything would have practical applications in everyday life is more difficult to predict. It's worth remembering that quantum mechanics gave us the semiconductor revolution and relativity led to nuclear power; neither of those breakthroughs could have been imagined just a few decades earlier. "It is rare that fundamental new theories do not have practical applications," says physicist Glenn Starkman of Case Western Reserve University in Cleveland. "I have no doubt that if string theory / M-theory proves to be the correct Theory of Everything, given enough time we will find that it allows us to build a better mousetrap—or, more likely, a better video-game display." More important, however, is the profound impact that such a theory would have—in the long term—on the human psyche. In time, it would lead to a new view of our place in the cosmos, just as the work of Copernicus and Darwin did in the past. Those ideas have made us humbler but wiser, and both met with resistance. A Theory of Everything may meet similar resistance. But when finally developed, it will give even deeper insights into our place in the natural world—insights which today we cannot imagine.

A Theory of Everything will, of course, *unify*: it will bring many seemingly disparate phenomena in particle physics and cosmology under one scientific "umbrella." But the final theory will also answer the question asked by Thales, Democritus, and Empedocles some 2500 years ago: "What is matter made of?" Democritus guessed the answer was "atoms." Scientists later discovered that atoms were made of protons, neutrons, and electrons, and later still that protons and neutrons were made of quarks. If the latest theories are correct, then the basic building blocks of the cosmos are actually unfathomably small loops of string, or, in the M-theory picture, some sort of membrane or higher-dimensional "brane."

These successive theories, from Democritus and the Presocratics to those being studied at Fermilab and CERN, have taken us deeper and deeper into the heart of matter. In one sense they have taken us closer to home; after all, they describe what we, as well as the universe, are made from. But at the same time they have described a world more and more remote from everyday experience—an alien landscape described only by mathematics.

For the Greeks, everything was made from earth, air, fire, and water; for us, perhaps, everything is made of strings a billion billion times smaller than an atom. The Greek view was based on speculation and philosophy; the modern view is supported by established science—though strings, for now, lie at the very edge of what science can reveal. Can either view be labeled as "true"? The Greek picture deals with things we know—things we can see, touch, and feel. The atom, thanks to devices like the scanning tunneling microscope, can be "seen" now too, though in a rather indirect manner. (This device records the feeble electric current that flows between the object being studied and an incredibly fine tungsten probe, itself just a few atoms wide.) As for strings and branes, there is no hope of studying them with any kind of conventional microscope. At best, we could study large strings by bouncing smaller strings off them—and even this would require a technology far beyond anything we can imagine. No matter how much indirect evidence points to their existence, we will never "see" strings in the way that we see fire and water.

For some people, this remoteness is unsettling; a few may even feel deeply alienated. This "physics angst" was described recently by historian Jacques Barzun. Modern physics, he says,

> deprived human beings of any object of cosmic contemplation. The actual order of the heavens and the workings of nature on earth were alike unimaginable—no poet could make an epic out of them, as Lucretius and Milton had done, or address a lyric to the moon.... None of the new terms coined at the scientific mint were evocative. *Electron, photon,* and later: *quark* ...which popularizers keep idiotically calling "building blocks" of the universe, carry no suggestion of being blocks. Even "particle" ...is a misnomer, since its instant-flash existence leaves but a dot on a sensitive plate; it never flies into one's eye and makes it water.

*"Scientists confirmed today that everything we know about the
structure of the universe is wrongedy-wrong-wrong."*

I sympathize with Barzun, and I'll try not to take his put-down
of science popularizers as a personal attack (in this book I have
indeed been guilty of "idiotically" referring to the subatomic
"building blocks" of nature). But he goes too far when he says that,
thanks to physics, poets can no longer be awed by the natural
world. (Perhaps he should read Richard Ryan's "Galaxy," A.M.
Sullivan's "Atomic Architecture," or John Haines' "Little Cosmic
Dust Poem.") Still, the sense of cosmic alienation he describes is
no doubt shared by many ordinary people who feel that modern
science doesn't describe *their* world.

For those who were already straining to comprehend Einstein's
curved space-time or Heisenberg's subatomic world of inherent
uncertainty, a world made of tiny strings 10^{-33} centimeters across
is simply absurd. For some, the quark is *already* absurd. Many
philosophers—and even a few scientists—ask whether it is meaningful to say that such objects "exist." As science carries us farther

and farther from the world of our sensory experience, we are forced to ask: "What is real?"

Models vs. Reality

Try to decide if you believe the following statements or not:

- You are holding a book in your hands right now.

- World War II happened.

- Antarctica exists.

- Dinosaurs once lived.

- Nuclear reactions power the stars.

- A collapsing star can form a black hole.

- Matter is made of atoms, which are made of quarks and electrons.

- Quarks and electrons are made of tiny strings or branes.

The point of this exercise is that the "truth" of statements like these falls on a continuum from the "certain" to the "uncertain" (in my opinion, only the last one is open to any serious doubt). In between there is a range of certainty, based on an accumulation of indirect evidence. (If you were born after 1945 and have never travelled to Antarctica, you have to rely on a kind of indirect evidence even for the second and third items—though I hope you'll agree they're indeed quite certain.) Advances in modern particle physics, like those reflected in the last two items on the list, seem to be especially vulnerable to doubt and uncertainty. Philosophers and skeptics of various stripes have spent their careers pondering the nature of the evidence and the degree of certainty, if any, that should be attached to such ideas.

Their struggle is not new. In the fourth century B.C., the Greek philosopher Plato described his view of reality by means of an allegory. Imagine, he said, a group of prisoners chained within a deep cave, facing the rear wall. A fire burns at the cave entrance and various objects stand between the fire and the prisoners. They cannot see the objects themselves, only the shadows of those objects cast on the cave wall. Perhaps, if they study those shadows carefully, the prisoners can make educated guesses about the nature of the objects—but they can never know their true forms.

Are scientists like Plato's cave-dwellers, guessing at a reality that remains forever unknowable? Over the centuries, philosophers have put forward endless variations on that idea. In the eighteenth century, George Berkeley, an Irish bishop and philosopher, went so far as to state that material objects do not exist unless they're perceived by a conscious mind. For Berkeley, there was no "real world," only the sensory impressions that make us infer such a reality. The English essayist and lexicographer Samuel Johnson famously denounced Berkeley's philosophy by kicking a large stone: "I refute it *thus*," he declared. Yet the work of Rutherford showed us that the atom—and thus the stone—is mostly empty space. As the astronomer Arthur Stanley Eddington lamented, "what Rutherford has left us of the large stone is scarcely worth kicking."

You don't have to be anti-science to wonder about the "existence" of atoms. Neither Berkeley nor Eddington was an enemy of science, and neither was physicist Niels Bohr—one of the founders of quantum theory—when he proclaimed that "The quantum world does not actually exist." In his view, we have only a *description* of the quantum world, expressed by a set of mathematical equations. "It is a mistake to believe that the purpose of physics is to find out how nature is made," he said. "Physics is interested only in what we can *say* about nature" (my italics).

Science-fiction writers often tell stories in which the main character is shrunk down to the size of a germ, a molecule, or even an atom. At first glance, this may seem to help us settle the matter: "Shrink me down to the size of an atom," the hopeful investigator says. "*Then* I will see is what atoms are really like!" Unfortunately, the quest is futile. In order to see an object, light must reflect off the object and enter our eyes. To see an object in detail, the size (wavelength) of the light waves must be smaller than the object being studied. A typical atom, however, is about 5000 times smaller than the shortest wavelength of light that our eyes can perceive. If you were reduced to the size of an atom, you would not "see" anything at all.

Although scientists have learned a great deal about the atomic world, it is fruitless to ask for a definitive description. The situation is explained eloquently by physicist and science writer John Gribbin:

> The point is that not only do we not know what an atom is 'really,' we *cannot* ever know what an atom is 'really.' We can only know what an atom is *like*. By probing it in certain ways, we find that ...it is 'like' a billiard ball. Probe it another way and we find that it is 'like' the Solar System. Ask a third set of questions, and the answer we get is that it is 'like' a positively charged nucleus surrounded by a fuzzy cloud of electrons. These are all images that we carry over from the everyday world to build up a picture of what the atom 'is.' We construct a model, or an image; but then, all too often, we forget what we have done, and we confuse the image with reality.

One might hear echoes of Plato in that passage, and perhaps there's nothing wrong with that. Gribbin does not say that atoms don't exist—only that it is meaningless to ask for a definitive account of an atom's properties; all we have are the descriptions given by our models. Yet there are still those who, like Berkeley, seem to believe there is *only* the model and nothing else.

By now we're getting into philosophy, an area where most physicists have enough sense to tread lightly. Indeed, a quick perusal of the philosophy-of-science literature suggests that even the philosophers haven't quite decided what to call the various viewpoints described above. If you side with Samuel Johnson—roughly: "There is a real world, and our most successful models give us a glimpse of that reality"—then you subscribe to *realism*. And you're in good company, alongside Kepler, Galileo, and Einstein. If you side with Berkeley—"There are only sensory impressions, from which we build models"—then you can spend a few days choosing a label: you might be a *positivist, empiricist, phenomenalist, instrumentalist, operationalist, fictionalist, subjectivist*, or even a *constructivist*. (And they say physics is confusing!) If you take that side, you're again in good company. This is where we find Niels Bohr and Stephen Hawking, along with philosophers such as David Hume, Immanuel Kant, and John Stuart Mill.

Of course, many physicists (perhaps most) are only vaguely interested in such issues. Ed Witten, the string theorist, gave a simple response that I suspect many physicists would echo: "I'm not going to identify myself on that [philosophical] spectrum. I'm not going to identify myself as being any kind of 'ist.'" Yet there are a few scientists, like Stephen Hawking, who make their particular philosophy almost a point of pride. Referring to the 11-dimensional world of string theory, Hawking says: "...as I am a positivist, the question 'Do extra dimensions really exist?' has no meaning. All one can ask is whether mathematical models with extra dimensions provide a good description of the universe." And later: "From the point of view of positivist philosophy ...one cannot determine what is real. All one can do is find which mathematical models describe the universe we live in."

It's worth taking a brief aside here to mention the most peculiar breed of anti-realist, the "social constructivist." These are people who believe that science is nothing but a set of beliefs that stem from the cultures of the scientists who dream them up. In other words, we don't *discover* quarks and quasars, we *invent* them. True, scientists grow up surrounded by cultural influences just like the rest of us—but that does not mean that the discoveries themselves are social constructs. As the philosopher James Robert Brown puts it, "point of view matters to the direction of research, but not to the facts themselves." As Brown points out, the social constructivists are guided by a sense of equality—"all theories are equal"—but always wind up supporting certain views (like those of native groups) and rejecting others (like those of fundamentalist Christians who support a literal interpretation of Genesis). "The double standard," says Brown, "is glaring." Indeed, today's social constructivists would probably want to distance themselves from those enemies of Einstein who saw his theory of relativity as a construction; the most hateful among them dismissed his work as "Jewish science." As Paul Davies says:

> These days it's fashionable in some university arts departments to present science as just a set of myths—as some sort of cultural phenomenon that is to be put alongside ancient wisdom, folklore, and other systems of thought as equally valid. In other words, it's just a part of our culture; it doesn't really represent anything that's really 'out there.' I think that's absolute twaddle. It is perfectly clear to me that the universe is ordered in a rational, mathematical way, and that in doing science we uncover that order—we read that order out of nature, we don't read it into nature.

When the evidence for some particular theory is overwhelming—"you are holding a book in your hands"—you don't bother with a model. In determining whether stars are powered by nuclear reactions, the evidence is more indirect and the model becomes indispensable. If the model proves successful in describing stellar physics in detail, we conclude that stars *really are* powered by nuclear reactions.

And what of atoms and quarks? Most physicists, I suspect, would say there really are such entities in nature, even if our descriptions of them, based on our best available models, are incomplete. The tiny strings and membranes posited by string theory and M-theory are more speculative, but they appear to at least hint at structures that exist in nature—of course using the word "structures" in a very loose sense. Yes, at times modern physics borders on pure mathematics; at other times it touches on philosophy. But, as Einstein asked, "Why should anybody go to the trouble of gazing at the stars if he did not believe the stars were really there?"

What About God?

SCIENCE FINDS GOD

COVER STORY IN *NEWSWEEK*, 20 JULY 1998

Physics isn't a religion. If it were, we'd have a much easier time raising money.

LEON LEDERMAN

So far I've said quite a bit about philosophy and mathematics, but very little about religion. And yet the frontiers of physics—including the search for a Theory of Everything—inevitably draw one toward a region where science begins to encroach on matters of theology. How did the universe begin? Where did the stars, planets, and galaxies come from? Could there be intelligent life on other worlds? At one time, only theologians and philosophers would have asked such questions. Today, astronomers, physicists, and cosmologists routinely investigate these matters, without recourse to the divine. The "God question" remains thorny, frequently emotional, and—judging by the number of books and magazine articles being written on the subject—as popular and provocative as ever. I can't claim to offer any startling new insights, but a review of what scientists and scholars have said over the centuries—and a look at what they are saying today—might be helpful.

Our story began with the birth of science, from the Greeks to the Scientific Revolution. The beliefs of the ancient Greeks covered a broad spectrum; some, like the Presocratics, were materialists who rejected anything supernatural; others, like Plato and Aristotle, believed in a higher power that was responsible for creating an ordered, intelligible universe. In the case of the great Renaissance scientists—Copernicus, Kepler, Galileo, Newton— the situation is much more clear-cut: there is no doubt that they were all men of deep and unquestioning faith. All of them found a "role for God"—if I can put it in such simple terms—within their scientific picture of the cosmos. Copernicus, for example, credits gravity to "the divine providence of the Creator." Kepler felt he was gaining a deeper appreciation of God's handiwork through his astronomy: "Our worship is all the more deep," he said, "the more clearly we recognize the creation and its greatness." In their faith, we hear echoes of the Bible—especially passages like the opening words of Psalm 19: "The heavens declare the glory of God; and the firmament showeth his handiwork."

The Scientific Revolution had not eliminated God. Rather, by declaring the universe "knowable," it found a new place for mankind within that universe—a place from which God's creation could be not only seen but understood. The British scientist Mary Somerville, who wrote popular accounts of the Newtonian world in the early nineteenth century, believed that God "endowed man with faculties by which he can not only appreciate the magnificence of His works, but trace, with precision, the operation of His laws, use the globe he inhabits as a base to measure the magnitude and distance of the sun and planets ...the first step of a scale by which he may ascend to the starry firmament."

An important shift, however, came in the second half of the nineteenth century, particularly with the work of Charles Darwin (1809–82). With his theory of evolution by natural selection, Darwin showed that complex living things could evolve from simpler organisms without divine intervention. Geologists, meanwhile, had uncovered the slow but steady processes that shaped the continents and oceans over millions of years—with human beings obvious latecomers to planet Earth. Astronomers, in turn, were discovering the vastness of the universe: by the 1920s, it was clear that our solar system was embedded in an immense, spiral agglomeration of stars, the Milky Way, and that ours was just one of a staggeringly large and possibly infinite number of such galaxies. The universe, the earth, living creatures—all seemed to evolve according to seemingly impersonal, uncaring laws.

True, there are puzzles that science has not yet solved: the origin of the universe, the origin of life, the origin of consciousness. Perhaps science will never answer these questions to the satisfaction of all. There are some who would invoke God to explain these gaps in human knowledge—"if science can't explain it, God must have done it"—what philosophers call a "God of the gaps" argument. For those who long for the traditional religious experience, however, such reasoning is hardly satisfying. For one thing, it suggests a rather limited role for God; for another, it pins theology to "current science," forcing one to alter one's beliefs with each successive scientific advance.

Though Einstein frequently referred to "God" in his writings, he had no room for a "God of the gaps" and even less room for a God who intervenes in human affairs. Those who wish to portray Einstein as a religious believer often quote his famous aphorism, "Science without religion is lame, religion without science is blind." But a closer look at his convictions shows he was far from being "religious" in the traditional sense. "My comprehension of God," he once said, "comes from the deeply felt conviction of a superior intelligence that reveals itself in the knowable world," a God "who reveals himself in all that exists...." If one had to label such a feeling, he said, it could be called *pantheism*—roughly, the identification of God with nature. (This was the outlook of the seventeenth-century Jewish philosopher Baruch Spinoza, whom Einstein greatly admired.) What Einstein did *not* believe in was a "personal" God, a deity who "concerns himself with the fate and actions of human beings."

Even with the advances of the twentieth century, religious believers were still able to find echoes of their sacred texts in the new discoveries of science. In cosmology, the big bang model reminded many of Biblical creation. "Let there be light," it seemed, could apply to both descriptions of the origin of the universe. (Why God chose to create the universe at that moment, and how he occupied his time for the next 15 billion years, was never quite as clear.) Meanwhile, the development of quantum theory offered a whole new canvas for those who wished to paint science as some sort of spiritual quest; numerous popular writers tried to link quantum mechanics with Eastern mysticism. None of the pioneers of quantum theory, however, felt there was such a link, and neither do most physicists today. Books like *The Tao of Physics* and *The Dancing Wu Li Masters* contain "some good physics writing," says particle physicist Leon Lederman, but "the authors jump from solid, proven concepts in science to concepts that are outside of physics and to which the logical bridge is extremely shaky or non-existent."

By the second half of the twentieth century, atheist scientists were beginning to speak up—among them Richard Feynman, who said the idea of a universe created "simply as a stage for God to watch man's struggle for good and evil seems to be inadequate."

(When he expressed a similar sentiment in a TV interview in 1959, a California station refused to air the item.) These days, the most outspoken non-believer (at least among physicists) is Steven Weinberg. In his 1977 book *The First Three Minutes*, one sentence wound up being quoted far more than any other: "The more the universe seems comprehensible," he wrote, "the more it also seems pointless." Fifteen years later, in *Dreams of a Final Theory*, he added that "it is hardly possible not to wonder whether we will find any answer to our deepest questions, any sign of the workings of an interested God, in a final theory. I think that we will not." Just for good measure, he later wrote in the *New York Review of Books* that the ultimate laws of physics "will be quite impersonal, not showing any sign of concern for human beings."

Of course, not all of today's scientists are atheists. In 1999, the American Association for the Advancement of Science (AAAS) organized a symposium on science and religion; the meeting, held in Washington, drew hundreds of researchers who, in many cases, have no difficulty in embracing both science and faith. They included Joel Primack, a cosmologist who holds a deep interest in Jewish mystical traditions; John Polkinghorne, a particle physicist who became an Anglican priest; and Owen Gingerich, an astronomer who believes in a "powerful Creator with a plan and an intention for the existence of the entire cosmos." Abdus Salam, who shared the 1979 Nobel prize with Weinberg and Sheldon Glashow, came from a religious Muslim background; Islam, he said, encourages the study of nature and presents no conflict with science. Even the notion of a more traditional "personal" God is very much alive; in a recent survey, 40 per cent of U.S. scientists said they believe in a God who responds to prayer and rewards the faithful with life after death. However, among scientists at the top of their profession, the proportion seems to be smaller. A recent survey found that only seven per cent of members of the prestigious National Academy of Sciences believe in a personal God.

For many, though, the idea of a personal God—one who cares about human affairs—is difficult to reconcile with what science has told us about the universe. Some, like Paul Davies, have embraced a less tangible deity. "I think of God—and I use the word with great caution—as the architect and guarantor of this magnificent order in the universe," he says. The alternative, says Davies, is to return to a God of the gaps—a God who merely "moves atoms around in competition with other forces of nature." That sort of God, he says, is so unsatisfying it would be little better than no God at all. How much more wonderful, Davies says, to envision God

> timelessly guaranteeing a set of laws that are so ingenious that without any crude, divine meddling or intervention, without any pre-existing design which is implemented by prodding and poking and manipulating, can bring into being such magnificent things as human beings with their power to reflect upon the universe....That's a much more magnificent picture than the crude fairy story, cosmic-magician God, which I wish we could abolish once and for all.

Indeed, many theologians have come to see God in this light as well. Just as science has evolved since the time of Newton, so theology has evolved since the time of Augustine. "There are Christian theologians who have very abstract notions of God, and who would be much more comfortable with definitions that are somehow non-personalized," says Jensine Andresen, a professor of theology at Boston University. Many who are embracing this more abstract idea of God, she says, are doing so "because of conversation with science and what they learn about science."

For atheists like Weinberg, however, such ideas hold little sway. Personal or abstract, knowable or mystical—any God is one too many. "One of the great achievements of science," says Weinberg, "has been, if not to make it impossible for intelligent people to be religious, then at least to make it possible for them not to be religious. We should not retreat from that accomplishment."

Believing in God as a matter of faith is one thing; trying to invoke science to support such a belief is quite a different matter. What we hear less of today than we did in the time of Copernicus and Kepler (notwithstanding the *Newsweek* headline quoted earlier) is scientists who claim to have seen God, or specific evidence for God, in nature. True, there are still physicists who manage to work God—usually metaphorically—into their pronouncements on nature. Stephen Hawking, for example, famously said that learning the ultimate laws of physics would allow us to "know the mind of God." (He would later explain that he does not believe in God, unless one defines God as the embodiment of the laws of physics.) A similar I've-seen-God brouhaha erupted in 1992, when cosmologist George Smoot and his colleagues discovered tiny fluctuations in the cosmic microwave background radiation. In an off-the-cuff remark to reporters, Smoot—referring to a microwave image of deep space that resembled a giant blue and pink hamburger—said the discovery was "like looking at God." (He later issued a disclaimer, saying he would leave the religious implications to the theologians.) Perhaps such urges are natural when writing books and speaking to journalists. When scientists are engaged in their work, however, they usually leave any notions of a personal deity at the door. Science, in spite of the catchy metaphors, has become a secular pursuit.

Yet there are still those who try to twist science to satisfy their theology. As this book was going to press, proponents of "Intelligent Design" were trying to have their theory—supposedly an alternative to Darwinian evolution—taught in Ohio schools. Supporters of the theory claim that certain microbiological structures are too complex to have evolved through Darwin's process of natural selection (their favorite example is the "bacterial flagellum"—a propeller-like structure that certain bacteria use for propulsion). Instead, they claim an "intelligent agent" must be responsible. Because they'd like to see the theory taught in public schools, they're usually careful to avoid mentioning God. Occasionally, however, they tip their hand. William Dembski, one of the leading proponents of Intelligent Design, has written that "Christ is indispensable to any scientific theory" and that "all disciplines find their completion in Christ and cannot be understood apart from Christ." (A thorough scientific critique of Intelligent Design has been given by biologist Kenneth Miller— himself a practicing Catholic—in his recent book, *Finding Darwin's God.*)

The Improbable Universe

In cosmology, meanwhile, there are those who see evidence for "design" in the heavens—particularly in what they see as the staggering improbability of a cosmos that would end up being hospitable to intelligent life. The dilemma centers on the values of the fundamental constants that govern the physical world. If the strength of gravity were slightly weaker, for example, or if the charge on an electron slightly different, then galaxies, stars, and planets could not have formed, nor could the complex chemicals needed for life. The result would be a dead universe. (The argument is sometimes called the "anthropic principle"; because that term has been used by different writers to mean different things, however, I prefer to simply call it the "fine-tuning problem.") Were we just lucky? Perhaps—and some physicists do dismiss it as sheer luck. Remember, though, words like "improbable" and "lucky" lose much of their meaning when we only have a single example—universe—in front of us. After all, if the cosmos had *not* bee hospitable to life, we wouldn't be here to argue about it. (At the AAAS symposium, Steven Weinberg was asked to address the question, "Was the universe designed?" He delivered a 90-minute lecture with the simple title, "No.") Not everyone, however, is satisfied with the "we wouldn't be here otherwise" comeback. The philosopher Richard Swinburne uses the analogy of a prisoner facing a firing squad: suppose all the marksmen fire—and miss. The prisoner can say he wouldn't be alive *unless* they had all missed—but he might also seek an explanation for *why* they missed.

Some try to use the fine-tuning problem as a proof for the existence of God, though traditional believers may find little comfort in a God who merely "sets the controls for the universe" at the moment of the big bang and then sits back to watch the show. Indeed, in a universe billions of years old, stretching billions of light years across, it's hard to imagine that human beings are in any way "central." As writer Timothy Ferris puts it, "The larger the universe looms, the sillier it becomes to maintain that it was all put together for us." Nearly 70 years ago, Bertrand Russell mocked the idea of "cosmic purpose" with his typical caustic wit: "If I were granted omnipotence, and millions of years to experiment in, I should not think Man much to boast of as the final result of all my efforts."

Another approach to the fine-tuning problem has been suggested by cosmologists such as Andrei Linde of Stanford and Sir Martin Rees of Cambridge. The answer, they speculate, may come from the idea of the *multiverse*—the notion that the big bang that created our universe was only one of many such explosions. (The idea for the multiverse grew out of the inflation model of the big bang, which we heard about in the previous chapter. It is highly speculative—but no more so than the many-worlds interpretation of quantum mechanics or the hidden dimensions and parallel brane-worlds that we discussed earlier.) If there really are multiple iverses out there, then we have a way out of the fine-tuning dilemma: our universe may appear special, but it's actually just one of many universes, most of which are not so finely-tuned, and are therefore not home to microbes, mice, and men. If one sells enough tickets in the cosmic lottery, the analogy goes, someone is bound to win. The multiverse idea is, of course, far from being an established theory. "I accept that all these discussions about multiple universes [are] speculative," Rees says. "But I would simply claim they are speculative science, not merely metaphysics."

Science is not yet in a position to address the fine-tuning problem, because we simply do not know why those physical constants have the values that they have. As Rees says, "The status and scope of [these] arguments, in the long run, will depend on the character of the (still quite unknown) physical laws at the very deepest level." In other words, we need a Theory of Everything. As we've seen, a successful unified theory of physics will explain why those physical parameters—the strength of gravity and the other forces, the masses of the particles, and so on—have the values that we see in our universe. With this more complete picture, the "fine tuning" will no longer be quite so mysterious. We can still use the word "design," if we wish, so long as we use it carefully: The universe would turn out to be hospitable to intelligent life because the laws of nature imposed such a design.

Faith in a Final Theory

One God, one law, one element ...

To which the whole creation moves.

ALFRED, LORD TENNYSON

Science may not embrace religion to the degree that some would wish, but there may be other, perhaps more subtle links between faith and science. One could argue, for example, that all scientr regardless of their religious views, have to believe in an inherer order in nature that makes science possible. Paul Davies says that this faith—however we may label it—derives from religion. All scientists, he says, "really do believe there's something underpinning it all. And if you take that [entity] and give it the name 'God'—which is optional—then I think that [this idea of] God is very close to the God of classical [monotheistic] theology." Einstein once expressed a similar sentiment: "There is no doubt," he said, "that all but the crudest scientific work is based on a firm belief—akin to a religious feeling—in the rationality and comprehensibility of the world."

This "religious feeling" that drives science—this belief in a rational, ordered universe—is central to every branch of scientific inquiry. The search for laws of nature is meaningless unless one believes that nature is indeed governed by such laws. The hope of achieving a Theory of Everything, above all, may owe its origins— and perhaps even its primary motivation—to a belief in order and simplicity that began with religious feeling. Physicist and writer John Barrow says that it was precisely such faith that ignited the search for a final theory:

> I think our belief that there exists a Theory of Everything is at root religious. On the face of it, there's no reason why we shouldn't be content with a perfectly working theory of the universe that says there are four fundamental forces; they're all completely different, and each is described by its own self-contained theory. But we feel uneasy about that. Nobody feels content with such a description. They feel it's somehow superior to have one law—one God, as it were, rather than many—and one can see how deeply embedded within our psyches this inclination is, to try to unify things together, and to regard fewer explanations as being better than more explanations.

Particle physicist Abdus Salam was once asked if, as a Muslim, his religious beliefs guided his research. "I wouldn't say consciously," he replied. "But at the back of one's mind the unity implied by religious thought perhaps plays a role in one's thinking." (Just to be perfectly clear: I am not saying that science *is* a religion—only that the drive to seek out a unified theory may have come, in part, *from* religion.)

Of course, the fact that we *want* the universe to follow a single, unifying law does not mean the universe *does* obey such a law. "It's quite conceivable that the universe doesn't run along lines which are in tune with human intuition," says Barrow. "The universe ...might be chaotic, completely irrational on the whole, but, here and there, there might be pockets of rationality—rather like oases of order within some infinity of chaos....So it is quite possible that things are not only strange, but much stranger than we could ever imagine." Why, then, did we begin looking for simple laws in the first place? Perhaps, says Barrow, it made us better able to cope with the often overwhelming flood of sensory information that we take in from our environment. The more phenomena we classified as "known," the less of the "unknown" there was for us to fear. "It's part of the adaptive evolutionary process, perhaps, of understanding fully our environment, and integrating it into a single description."

If we pursue this line of reasoning further, we may end up linking the pursuit of science to the demands of our neurobiology—an idea that can be traced back to the eighteenth-century German philosopher Immanuel Kant. He argued that our minds impose a certain amount of structure on the raw data that we take in with our senses. As a result, Kant said, we inevitably draw conclusions about nature that in fact stem only from the way our minds organize information. (Modern neuroscientists are now addressing this very question, as they investigate the way in which the brain processes and organizes sensory information.)

For many scientists, such speculation about the origins of scientific thought—whether from the perspective of anthropology, history, or neuroscience—is beside the point. They search for a simple, unified picture because it is a strategy that works. "Physicists in the past who have looked for simplifying, unifying assumptions have done well. Better than physicists who haven't," says Sheldon Glashow. "But there's no *reason* that things get simpler. They could become more and more chaotic and more and more complicated." This possibility—that after probing beyond a certain point the laws of nature stop becoming simpler and start becoming more complex—strikes most physicists as absurd. And yet there is no "proof" that the laws of nature must continue to be simple, no matter how deeply one explores.

Einstein often spoke of the necessity of nature's laws being simple; the mathematical jargon notwithstanding, he said, all physical theories should be simple enough that even a child could understand them. He spoke on this subject again, perhaps for the last time, at a seminar on relativity in Princeton. It was the spring of 1954; he had one year left to live. Someone asked him how he would feel if the final laws of nature turned out *not* to be simple. "Then I would not be interested in them," he replied.

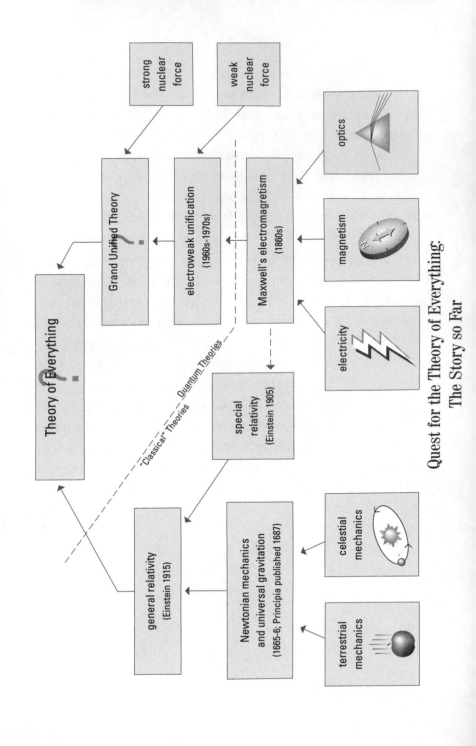

Quest for the Theory of Everything:
The Story so Far

Epilogue

The answer to the Great Question ... of Life, the Universe, and Everything ... is Forty-two.

—DOUGLAS ADAMS, *THE HITCH HIKER'S GUIDE TO THE GALAXY*

he quest for a unified theory of physics—the Theory of Everything—brings us to the very frontier of scientific nquiry. It forces us to ask how far science can take us. That's a question that scientists and philosophers have debated for centuries, with no resolution in sight. In the last few years, it's led to a flurry of articles, books, and speculation—triggered in part, no doubt, by the coming of a new millennium and the general urge to ask deep questions that such landmark dates seem to foster.

When science offers successful theories—as it has so many times in the last 400 years—it's tempting to say there may not be any limits at all; the farther we probe, the more we will discover. Yet such a vision may be unwarranted. Certainly, our ability to *explore* the universe has limitations. Because of the impossibility of faster-than-light travel, and because of the enormous drain on resources that interstellar travel would entail (even for a civilization far more advanced than our own), distant regions of the universe may remain eternally out of reach. Even getting *information* from distant regions is limited by this universal speed limit, as no signal of any kind can travel faster than light. The interiors of black holes, too, must remain forever off limits; no one who enters could exit or even send back a signal. Our ability to process information may be constrained also: our computers are getting faster and faster, but they too may reach a point beyond which they cannot pass, again limited by the speed at which information can flow.

Finally, our own biology may impose limits on what we can comprehend—after all, we have finite, one-and-a-half-pound brains, which evolved to help us survive on the African savannah and have changed little in the last 100,000 years. The fact that our brains turned out to be of use in solving differential equations is improbable enough, the argument goes; why should we be able to understand the deepest laws of nature as well? Other crea certainly have *their* limits; as many writers have pointed c teaching quantum mechanics to a dog is extremely difficult. W we eventually run up against such a barrier—a barrier set by the limitations of our own minds? There is no way to be certain, but most of the physicists I've spoken with say there is no sign that we are nearing such limits. "The empirical evidence is that graduate students today are as smart as they've ever been," says string theorist Ed Witten. "There might be limits [to human understanding] but there's no sign that we've reached them yet."

Reaching for the Grail

Suppose the universe is governed by a set of simple, elegant laws, and suppose that human beings are clever enough to deduce those laws. How will we know when we've accomplished that task? When do we turn from searching to celebrating? The difficulty comes from the nature of science itself: a scientific theory is never truly "final." The best one can hope for is that it encompasses more, and explains more, than theories that came before. In the words of the twentieth-century Austrian philosopher Karl Popper:

> Science never pursues the illusory aim of making its answers final, or even probable. Its advance is, rather, towards an infinite yet attainable aim: that of ever discovering new, deeper, and more general problems, and of subjecting our ever tentative answers to ever renewed and ever more rigorous tests.

Einstein expressed the same sentiment when he said that our theories are "hypothetical, never completely final, always subject to question and doubt." The problem is that a new theory, no matter how successful, can always be shown to have some limitation. One can never rule out the possibility that a new experiment

...be performed which will force the theory to be amended. As ...on Lederman explains:

> It's very hard to prove a theory is right. You go to the laboratory, you can prove a theory is wrong—that's easy. The theory makes a prediction and your experiment shows it's wrong. But the other result is "maybe it's right"....You can never prove it's right. All you can do is check that maybe it's right. And if you get enough 'maybes,' it becomes a part of your belief system.

So what, exactly, do we mean by a "Theory of Everything"? Perhaps we should not be so presumptuous as to call it the *final* theory. A better definition would involve the list we had at the start of the last chapter: a Theory of Everything, for our purposes, would be a theory of physics that explains the Standard Model of particle physics and—as an added bonus—explains puzzles like Einstein's cosmological constant and perhaps the nature of the big bang itself. This, as we've seen, is not "everything," but it certainly is a lot. And many of today's leading theorists feel it is an attainable goal.

True, it may take another Newton-calibre genius to put it all together, and no one can say how long we will have to wait. Yet at any moment, writes Steven Weinberg, we may find an article "by some previously unknown graduate student, laying it all out." It may not happen in this century, he cautions. "But I think it will happen, and when it does it will end a certain chapter in the history of science: the search for the fundamental principles that underlie everything." The physicist Lee Smolin predicts that we'll have "the basic framework of the quantum theory of gravity by 2010, 2015 at the outside." By the end of this century, he says, the theory "will be taught to high school students around the world."

When I asked Leon Lederman about the criteria for declaring the problem solved, he gave a concise answer: "Explain the Standard Model; explain the origin and evolution of the universe.

If you can do that, you're finished." Simple, right? Well, it *is* simple. "The weakness of particle physics," Lederman continues, "is that we only address one question: how does the universe work? And the Theory of Everything, by definition, is a theory which explains how the universe works. So we brush our clothes off and say, okay, this subject is finished. What's next?" To those who have not spent their lives immersed in the world of equations, theories, and particle accelerators, such confidence may sound surprising. It may even sound arrogant. "Yes," concedes Lederman, laughing. "I can't argue with that. I think it's the ultimate arrogance. But I think we've been arrogant for a very long time. Encouraged, if you like, by the fact that *science works*. It's had its blind alleys and its mistakes; it has its inefficiencies—but ultimately it's self-correcting, and has had a powerful effect on human capabilities, human potentialities, and human imagination."

Ed Witten shares Lederman's confidence, but stresses that we are still far from the goal. He compares the search for the ultimate laws of nature to the work of a chef peeling back the layers of an onion: sometimes you feel you're nearly there, only to discover another, deeper, layer. "Today, we know that we haven't reached the center of the onion, because the various bits and pieces that we know about aren't coherent enough—they're obviously derivatives of something more fundamental," Witten says. "If you ever reach the point where the pieces do fit together and you don't feel that compulsion to go on—then I guess you would stop, and leave it for later generations to figure out whether they could do better." He pauses, and then adds: "I'm not completely certain that it *is* a question that will prove to have a definitive answer. It might be a question that only has better and better answers, when you probe more deeply. But I hope there is a definitive answer. And if there is a definitive answer, I hope that we'll live to see it."

ver the past eight chapters I may have given the impression that every physicist is obsessed with finding the Holy Grail—the Theory of Everything. Most are not. Of the roughly 40,000 members of the American Physical Society, fewer than one-tenth specialize in elementary particles, fields, and gravitation, the areas directly linked to the search for a unified theory. Many physicists would probably have sided with Richard Feynman: "People say to me, 'Are you looking for the ultimate laws of physics?' No, I'm not, I'm just looking to find out more about the world and if it turns out there is a simple ultimate law which explains everything, so be it, that would be very nice to discover." For perhaps a few hundred researchers, however, the quest for the Theory of Everything is quite tangible. It may not be the reason they go to work each morning; usually they set their sights on smaller, more manageable problems. (These days, just checking the morning e-mail can take up a good part of the day.) But the lure of the unified theory is there, quietly focusing their thoughts and ideas. The Holy Grail beckons.

The quest has already spanned 2500 years, from the bold ideas of the ancient Greeks to the equally daring ideas of the string theorists of the twenty-first century. We've seen how scientists in every age have struggled to find unifying principles—to see the order and logic behind nature's diversity. Twenty-five centuries of searching, thinking, experimenting, and re-thinking. And that effort, as we've seen, has brought many successes. Enough successes, as Leon Lederman put it, to foster a kind of arrogance— and perhaps that arrogance is justified. Finding a unified theory of physics would be one of the great intellectual achievements of all time—perhaps *the* great achievement. The Theory of Everything would reach across the vast sweep of the cosmos, encom... more than any previous picture of the world. It would em... the smallest bits of matter and energy that we can possibly im... ine, as well as the most distant and ancient reaches of the univers... It would govern all the particles and forces, quarks and quasars, gluons and galaxies. It would be expressed simply, through the language of mathematics—perhaps simply enough to display on a T-shirt.

Yet I can't help thinking of Empedocles, sitting on a weathered rock by the blue waters of the Mediterranean, pondering the structure of his world. Head in hand, he contemplates the barren sands, the salty air, the comforting fire, the vast unexplored ocean. In proposing his theory of the elements, he was as bold and arrogant as any scientist of his time or ours. But he also had at least one fleeting twinge of modesty. "Narrow are the powers that are spread through the body, and many are the miseries that burst in, blunting thought," he once said. "Men behold in their span but a little part of life—then, swift to die, are carried off and fly away like smoke, persuaded of one thing only: that which each has chanced on as they are driven every way. Who, then, boasts that he has found the whole?"

Recommended Reading

Though I consulted numerous written sources in researching *Universe on a T-Shirt*, the following list highlights those books, articles, and websites that I think will be most useful to the general reader. Most of the books are in print and readily available; a few of the older ones are out of print but can be easily found in libraries or purchased via the Internet.

History of Physics and Astronomy

Timothy Ferris's *Coming of Age in the Milky Way* (Anchor Books 1988) is a must-read for anyone interested in the story of humanity's struggle to comprehend the cosmos. To hear from the scientists in their own words, two recent anthologies are invaluable: *The World Treasury of Physics, Astronomy, and Mathematics*, edited by Timothy Ferris (Little, Brown and Company 1991) and *The Book of the Cosmos: Imagining the Universe from Heraclitus to Hawking*, edited by Dennis Richard Danielson (Perseus Publishing 2000). In spite of its age, *The Growth of Physical Science* by Sir James Jeans (Cambridge University Press 1947; Premier Books 1958) is still an entertaining and informative journey through the last 5000 years of scientific inquiry.

For the Presocratics and the early Greek philosophers, a recent popular account is *Early Greek Philosophy* by Jonathan Barnes (Penguin 1987). Two lively texts that carry the story up to the beginning of the Scientific Revolution are David C. Lindberg's *The Beginnings of Western Science* (University of Chicago Press 1992) and Thomas Goldstein's *The Dawn of Modern Science* (Houghton Mifflin Company 1980); the two are fairly similar in style and content.

The Scientific Revolution

A good introduction is I. Bernard Cohen's *The Birth of a New Physics* (W.W. Norton and Company 1985); Cohen also has a more thorough treatment, *Revolution in Science* (The Belknap Press of Harvard University Press 1985). Though over 50 years old, Angus Armitage's *The World of Copernicus* (Mentor Books, 1951) is still an enjoyable account of the great scientist's life and work.

For Tycho Brahe, the reader can choose between Victor E. Thoren's *The Lord of Uraniborg: A Biography of Tycho Brahe* (Cambridge University Press 1990) and John Robert Christianson's *On Tycho's Island: Tycho Brahe and his Assistants, 1570-1601* (Cambridge University Press 2000). For Johannes Kepler, a good starting place is still Carola Baumgardt's *Johannes Kepler: Life and Letters* (Philosophical Library 1951).

alileo, two good introductions are James Reston's *Galileo: A Life* .rperCollins 1994) and Stillman Drake's somewhat older volume, *Galileo* (Oxford iiversity Press 1980). To read the scientist's own account of his findings, see Jrake's *Discoveries and Opinions of Galileo* (Anchor Books 1957 and subsequent editions). For Newton, the most thorough treatment is Richard Westfall's *Never at Rest: A Biography of Isaac Newton* (Cambridge University Press 1980); a shorter version was published as *The Life of Isaac Newton* (Cambridge 1994). Also highly readable is Michael White's *Isaac Newton: The Last Sorcerer* (Perseus Books 1997).

A good overview of the work of Oersted, Faraday, and Maxwell can be found in Robert D. Purrington's *Physics in the Nineteenth Century* (Rutgers University Press 1997). Ivan Tolstoy's *James Clerk Maxwell: A Biography* (Canongate, 1981) is concise and enjoyable.

Einstein and Relativity

There are so many Einstein biographies on the market it's hard to pick a favorite. The one I consulted most often was *Albert Einstein* by Albrecht Fölsing (Penguin Books 1998); the most complete treatment of his scientific work is "*Subtle is the Lord...*": *The Science and the Life of Albert Einstein* by Abraham Pais (Oxford University Press 1982). To hear Einstein's own views on science, politics, and religion, his *Ideas and Opinions* (Dell Publishing Co. 1954 and subsequent editions) is indispensable. For an entertaining history of special relativity, see $E=mc^2$: *A Biography of the World's Most Famous Equation* by David Bodanis (Walker and Company 2000).

Quantum Theory

John Gribbin does an admirable job of making sense of quantum theory for the average reader in *In Search of Schrödinger's Cat: Quantum Physics and Reality* (Bantam Books 1984); another excellent introduction is *The Ghost in the Atom* by Paul Davies and J.R. Brown (Cambridge University Press 1986; Canto 1993). For a more detailed historical account, see *Inward Bound: Of Matter and Forces in the Physical World* by Abraham Pais (Oxford University Press 1986) or *The Second Creation: Makers of the Revolutions in Twentieth-Century Physics* by Robert P. Crease and Charles C. Mann (MacMillan Publishing Company 1986). Several useful and quite readable journal articles marked the one-hundredth anniversary of quantum theory, including "100 years of quantum mysteries" by Max Tegmark and John Wheeler in *Scientific American* (February 2001) and "One hundred years of quantum physics" by Daniel Kleppner and Roman Jackiw in *Science* (vol. 289, issue 548, 2000).

Modern Physics, Cosmology, and String Theory

A number of excellent accounts of modern physics and cosmology have appeared in the last few years. Two of the best, focusing primarily on cosmology, are *The Whole Shebang: A State-of-the-Universe(s) Report* by Timothy Ferris (Simon & Schuster

1998) and *Before the Beginning: Our Universe and Others* by Martin Rees (He
Books 1997). Focusing more on physics and equally worthwhile is Steven Weir.
Dreams of a Final Theory: The Scientist's Search for the Ultimate Laws of Nature
(Random House 1992).

The most thorough of the recent books on string theory is Brian Greene's *The
Elegant Universe* (W.W. Norton and Company 1999), although readers should be
warned that it is often highly technical. For a short introduction to string theory
and M-theory, see "The theory formerly known as strings" by Michael J. Duff in
Scientific American (February 1998), and on the Internet,
www.superstringtheory.com. For alternatives to string theory in unifying physics,
see Lee Smolin's *Three Roads to Quantum Gravity* (Weidenfeld & Nicholson 2000;
Basic Books 2001). Also worth reading is James McAllister's recent article, "Is beauty
a sign of truth in scientific theories?" in *American Scientist* (March-April 1998).

Philosophical and Religious Implications of Modern Physics

All of the books on "science and religion" seem to be heavily skewed by the biases of
their authors—and, I confess, so is the list I present here. Steven Weinberg touches
on the religion issue briefly but effectively in his book *Dreams of a Final Theory*,
mentioned above. Books from scientists offering a more sympathetic approach to
religion include *The Mind of God* by Paul Davies (Simon & Schuster 1992) and
Skeptics and True Believers by Chet Raymo (Doubleday 1998); either one would
serve as a good introduction to the field. For a philosopher's perspective, see James
Robert Brown's *Who Rules in Science? An Opinionated Guide to the Wars* (Harvard
University Press 2001). On quantum theory and philosophy, see John Gribbin's
Schrödinger's Kittens and the Search for Reality (Little, Brown, and Company 1995).
On the limits of science, see John D. Barrow's *Impossibility: The Limits of Science and
the Science of Limits* (Oxford University Press 1998). Though nearly 75 years old,
Arthur Stanley Eddington's *The Nature of the Physical World* (Cambridge University
Press 1928) is still worth reading; though an astronomer, Eddington was also an
astute philosopher of science, and his book has a remarkably modern feel.

Endnotes

Introduction

p. 1 *My ambition is to live to see...*
Lederman, Leon with Dick Teresi, *The God Particle: If the Universe is the Answer, What is the Question?* New York: Bantam Doubleday Dell Publishing, 1994, 21.

p. 1 *The longed-for Theory of Everything...* Barrow, John D., *Theories of Everything: The Quest for Ultimate Explanation*. London: Vintage, 1992, 115.

Shadows and Light

p. 7 *Zeus, father of the Olympians...* quoted in Easterling, P.E. and B.M.W. Knox, *The Cambridge History of Classical Literature*. Cambridge: Cambridge University Press, 1986, 127.

p. 8 "...the Lydians and the Medes..." Herodotus, *The Histories*, trans. Selincourt. New York: Penguin, 1996, Book I, part 74, 30.

p. 11 "inherited from their nomadic forbears..." Goldstein, Thomas, *The Dawn of Modern Science*. Boston: Houghton Mifflin Co., 1980, 47.

p. 12-13 "They saw the world as something ordered..." Barnes, Jonathan, *Early Greek Philosophy*. London: Penguin, 1987, 16.

p. 13 "first founder of this kind of philosophy" Aristotle, *Metaphysics*. quoted in W.K.C. Guthrie, *History of Greek Philosophy*, Vol. I. Cambridge: Cambridge University Press, 1962, 48.

p. 15 "An immortal god, no longer mortal..." quoted in Barnes, 192.

p. 16 "...men and women, and birds and beasts..." quoted in Lambridis, Helle, *Empedocles*. Tuscaloosa: The University of Alabama Press, 1976, 48.

p. 17 "Nothing happens in vain..." quoted in Barnes, 243.

p. 18 "By convention color..." quoted in de Santillana, Georgio, *The Origins of Scientific Thought*. New York: The New American Library, 1961, 145.

p. 18 "The atoms have all sorts of shapes..." quoted in de Santillana, 146.

p. 20 "If you look back at the earliest myths..." Barrow, John, Author interview for CBC Radio. 13 May 1997.

p. 21 "...unique inventions, never duplicated..." Cromer, Alan, *Uncommon Sense: The Heretical Nature of Science*. New York: Oxford University Press, 1993, 99–100.

p. 23 *...we must regard them at least as protoscientists...* Long, A.A. (ed.), *The Cambridge Companion to Early Greek Philosophy*. Cambridge: Cambridge University Press, 1999, 63.

p. 23 "Matter is constituted of particles..." Schrödinger, Erwin, *Science and Humanism: Physics in our Time*. Oxford: Canto, 1996, 11.

p. 23 "If their attempts sometimes look comic..." Barnes, 18.

p. 24 "the principal ingredients of a scientific approach..." Pullman, nard, *The Atom in the History of uman Thought.* Oxford: Oxford University Press, 1998.

A New Vision

p. 25 *It is clear that the earth does not move...* Aristotle, *On the Heavens.* In Barnes, Jonathan (ed.), *The Complete Works of Aristotle.* Vol. 1. Princeton: Princeton University Press, 1984, 487.

p. 26-27 **"...not necessary to probe into the nature of things..."** Goldstein, Thomas, *Dawn of Modern Science.* Boston: Houghton Mifflin Co., 1980, 57.

p. 27 **"no room for scientific observation..."** Goldstein, 55–56.

p. 29 *In expounding Scripture...* Lindberg, David C., *The Beginnings of Western Science.* Chicago: University of Chicago Press, 1992, 198.

p. 32 **"entities are not to be multiplied..."** quoted in Cohen, I. Bernard, *The Birth of a New Physics.* New York: W.W. Norton and Co., 1985, 127.

p. 32 *...will clamor to have me shouted down.* Copernicus, Nicolaus, *De Revolutionibus Orbium Caelestium,* trans. Dennis Richard Danielson. In Danielson (ed.), *The Book of the Cosmos: Imagining the Universe from Heraclitus to Hawking.* Cambridge, MA: Perseus Publishing, 2000, 104.

p. 33 **six to nine million copies** Ferris, Timothy, *Coming of Age in the Milky Way.* New York: Anchor Books, 1988, 62.

p. 34 **"...would compose a monster, not a man"** Copernicus, quoted in Danielson, 106.

p. 35 **the stars must be very far away** Copernicus turned out to be right. We now know that the nearest star, Alpha Centauri, is about four light-years away, some 6,500 times more distant than Pluto, the farthest planet. Because the stars are so far away, stellar parallax remained undetected until 1838, when three different astronomers succeeded in measuring the tiny, apparent shift in the position of a nearby star resulting from the earth's motion around the sun. Germany's Frederick Wilhelm Bessel was the first to publish.

p. 36 **"...multiplying spheres almost ad infinitum"** Copernicus, in Danielson, 116.

p. 36 **deprived humanity of a "special place"** For a thorough discussion of this misconception, see Danielson, Dennis R., "The great Copernican cliché." *American Journal of Physics* 69 (10) (2001): 1029–1035.

p. 36 **400,000 times larger** Ferris, 68.

p. 36-37 **"They had to adapt to a moving earth..."** Gingerich, Owen, Personal interview. 18 December 1999.

p. 37 **"...the marvelous symmetry of the universe..."** Copernicus, in Danielson, 117.

p. 38 **"through the triple holes in [his] nose"** Ursus, Nicolaus, *De hypothesibus astronomicis tractatus.* Quoted in Jardine, Nicholas, *The Birth of History and Philosophy of Science: Kepler's 'A Defence of Tycho Against Ursus' with Essays on its Provenance and Significance.* Cambridge: Cambridge University Press, 1984, 35.

p. 38 **"I noticed that a new and unusual star..."** Brahe, Tycho, *De Stella*

Nova, trans. John H. Walden, in Danielson, 129.

p. 39 "...a star shining in the firmament itself..." Brahe, in Danielson, 131.

p. 39 "If you want to settle down on the island..." quoted in Christianson, John Robert, *On Tycho's Island: Tycho Brahe and his Assistants, 1570–1601*. Cambridge: Cambridge University Press, 2000, 22.

p. 41 "Holding his urine..." quoted in Thoren, Victor E., *The Lord of Uraniborg: A Biography of Tycho Brahe*. Cambridge: Cambridge University Press, 1990, 468–9.

p. 42 "For a long time I was restless..." Baumgardt, Carola, *Johannes Kepler: Life and Letters*. New York: Philosophical Library, 1951, 31.

p. 43 "...could easily assemble a whole volume..." Cohen, I. Bernard, *Revolution in Science*. Cambridge, MA: The Belknap Press of Harvard University Press, 1985, 127.

p. 43 "foolish little daughter..." Baumgardt, 27.

p. 44 "a man of great benevolence" Ibid., 64.

p. 44 "...by an inexorable fate" Ibid., 66.

p. 45 "the most acute thinker ever born" quoted in Baumgardt, 17.

p. 45-46 "How much inventive power..." Ibid., 11–12.

p. 46 "...filled with unbelievable delight..." Ibid., 121.

p. Heaven and Earth

p. 48 *O telescope, instrument of much knowledge...* quoted in Ferris, Timothy,

Coming of Age in the Milky Way. York: Anchor Books, 1988, 95.

p. 52 "undoubtedly false" Cohen, I. Bernard, *Revolution in Science*. Cambridge, MA: The Belknap Press of Harvard University Press, 1985, 140.

p. 52 "...the first great scientific publicity stunt" Lederman, Leon with Dick Teresi, *The God Particle: If the Universe is the Answer, What is the Question?* New York: Bantam Doubleday Dell Publishing, 1994, 73.

p. 52 it *could* have happened Drake, Stillman, *Galileo at Work: His Scientific Biography*. Chicago: University of Chicago Press, 1978, 415.

p. 52 similar falling-body experiments For a discussion of the Pisa experiment, see Adler, Carl G. and Byron L. Coulter, "Galileo and the Tower of Pisa experiment." *American Journal of Physics* 46 (3) (Mar. 1978): 199–201. Also Segre, Michael, "Galileo, Viviani and the Tower of Pisa." *Studies in the History and Philosophy of Science* 20 (4) (1989): 435–451. For a discussion of scientists who performed the experiment prior to Galileo, see Weiss, P., "Weights make haste: Lighter linger." Science News Online, 4 December 2001 <www.sciencenews.org/sn_arc00/12_18 _99b/fob7.htm>. Also "Scientific Urban Legends." Lock Haven University of Pennsylvania. 4 December 2001 <www.lhup.edu/~dsimanek/sciurban. htm> and Dauben, "Joseph W. Galileo: The Early Years." In *Galileo's Experiment at the Leaning Tower of Pisa*. Endex Engineering, Inc., 4 December 2001

...endex.com/gf/buildings/ltpisa/
...ws/physnews1.htm>

...3 *A host of other stars are perceived...* Drake, Stillman, *Discoveries and Opinions of Galileo.* New York: Anchor Books, 1957, 47.

p. 53 **in just 24 hours** Reston, James, *Galileo: A Life.* New York: HarperCollins, 1994, 88.

p. 55 **as far away as China** Drake, Stillman, *Discoveries and Opinions of Galileo,* 59.

p. 55 **"the most famous man in the world"** Reston, 204.

p. 55 **"offered them a thousand times"** quoted in Baumgardt, Carola, *Johannes Kepler: Life and Letters.* New York: Philosophical Library, 1951, 86.

p. 56 **read to him over dinner** Reston, 191.

p. 57-58 **"Then spake Joshua to the Lord..."** *Holy Bible,* King James Version, Joshua 10: 12–13.

p. 58 **"one might fall into error"** Galileo, *Letter to the Grand Duchess Christina.* In Stillman Drake, *Discoveries and Opinions of Galileo,* 181.

p. 58 **"...he stooped to their capacity..."** Ibid., 211.

p. 58 **borrowed from Caesar Cardinal Baronius** Ibid., 186; also Gingerich, Owen, "The Galileo affair." *Scientific American* (August 1982): 137.

p. 58-59 **"...the Copernican system was not really the issue"** Gingerich, Owen, "The Galileo affair," 137–8.

p. 59 **"...a local Italian imbroglio"** Gingerich, Owen, Personal interview. 18 December 1999.

p. 60 **"...fraught with particular personalities..."** Ibid.

p. 60 **"...defend his scientific beliefs"** Drake, Stillman, *Discoveries and Opinions of Galileo,* 145.

p. 60 **"question of personality, not principle"** Reston, 195.

p. 60 **"...far more than the heliocentric theory"** Rubbia, Carlo, "Galileo and the popularization of science." *Archives des Sciences* 46(3) (1993): 279.

p. 60-61 **Thomas Hobbes dropped by as well** Reston, 279.

p. 61 **"tragic mutual incomprehension"** quoted in Raymo, Chet, "Righting Galileo's 'wrong.'" *Sky & Telescope* (March 1993): 4.

p. 61 **"the face of Galileo haunts..."** Reston, 139.

p. 61-62 **"...written in the language of mathematics..."** Galileo, *The Assayer.* In Stillman Drake, *Discoveries and Opinions of Galileo,* 237–8.

p. 62 **"sober, silent, thinking lad"** quoted in Westfall, Richard, *The Life of Isaac Newton.* Cambridge: Cambridge University Press, 1994 ed., 13.

p. 62-63 **"Tho Sir Isaac was not so lusty..."** Ibid., 13–14.

p. 63 **"fit for nothing but the 'Versity"** Ibid., 18.

p. 63 **"...and his head scarcely combed"** Ibid., 63.

p. 64 **"...the prime of my age for invention..."** Ibid., 39.

p. 66 "...reciprocally as the squares of their distances..." Cohen, I. Bernard, *The Birth of a New Physics*. New York: W.W. Norton & Company, 1985, 165.

p. 66 *Join me in singing the praises of Newton...* Cohen, I. Bernard and Anne Whitman, *Isaac Newton—The Principia: A New Translation.* Berkeley: University of California Press, 1999, 380.

p. 68 "Let Mortals rejoice..." quoted in Westfall, 313.

p. 68 **138 volumes on alchemy** White, Michael, *Isaac Newton: The Last Sorcerer.* Reading, MA: Perseus Books, 1997, 119.

p. 69 "Nature does nothing in vain..." quoted in Cohen, I. Bernard, *The Birth of a New Physics*, 127.

Flashes of Insight

p. 71 *...there exists a great unity in nature...* Walford, David, trans. and ed., *The Cambridge Edition of the Works of Immanuel Kant.* Cambridge: Cambridge University Press, 1992, 155.

p. 73 **lightning struck the kitchen** Tolstoy, Ivan, *James Clerk Maxwell: A Biography.* Edinburgh: Canongate, 1981, 112.

p. 75 "Galvanism...will lead to great discoveries" quoted in Dibner, Bern, *Oersted and the Discovery of Electromagnetism.* New York: Blaisdell Publishing Co., 1962, 4.

p. 76 "...completed with such rapidity" Ibid., 38.

p. 77 "Spirit and nature are one..." quoted in Gillespie, Charles Coulston, ed., *Dictionary of Scientific Biography.*

New York: Charles Scribner's Sons, 1981, 185.

p. 77 "Convert magnetism into electricity!" quoted in Tanford, Charles and Jacqueline Reynolds, "Voyage of discovery." *New Scientist* 19 (26) December 1992): 53.

p. 80 "...but facts were important to me." quoted in Ginzburg, Benjamin, *The Adventure of Science.* New York: Simon and Schuster, 1930, 229.

p. 81 "...by a child like that" quoted in Tolstoy, 14.

p. 81 "No jokes of any kind..." Ibid., 81.

p. 82 "...the subject stood it so well" Ibid., 75.

p. 83 "first of all, a synthesis..." Ibid., 126.

p. 85 "...light consists in the transverse undulations..." Maxwell, James Clerk, "On physical lines of force: Part III." *Philosophical Magazine* 4 (151) (1862): 22.

Relativity, Space, and Time

p. 88 *Absolute, true, and mathematical time...* Cohen, I. Bernard and Anne Whitman, *Isaac Newton—The Principia: A New Translation.* Berkeley: University of California Press, 1999, 408.

p. 88 *Henceforth space on its own...* quoted in Fölsing, Albrecht, *Albert Einstein.* New York: Penguin Books, 1998, 189.

p. 89 "...destroyed all illusions and ideals" Yehudi Menuhin, University of Adelaide website, 9 February 2002

ww.arch.adelaide.edu.au/~twyeld/
/cutsd/week3>

p. 89 **For historian Eric Hobsbawm**
Hobsbawm, Eric, *The Age of Extremes:
The Short Twentieth Century
1914–1991.* London: Little, Brown and
Company, 1994.

p. 91 **"We sail forth from the
harbor..."** quoted in Danielson, D.R.
(ed.), *The Book of the Cosmos:
Imagining the Universe from Heraclitus
to Hawking.* Cambridge, MA: Perseus
Publishing, 2000, 115.

p. 94 **"The patent job agreed..."**
Fölsing, 103.

p. 94 **"...to extract simplicity out of
complexity?"** Wheeler, John A. *Albert
Einstein: A Biographical Memoir*
(National Academy Press, 1980) as
excerpted in Ferris, Timothy (ed.), *The
World Treasury of Physics, Astronomy,
and Mathematics.* New York: Little,
Brown and Company, 1991, 568, 570.

p. 95 *...has brought about a
revolution...* quoted in Fölsing, 271.

p. 101 **by a factor of more than 400**
Bodanis, David, *E = mc²: A Biography
of the World's Most Famous Equation.*
New York: Walker and Company, 2000,
52.

p. 101 **In 1971, relativity was tested**
Davies, Paul, *About Time: Einstein's
Unfinished Revolution.* New York:
Penguin Books, 1995, 57.

p. 103 **"...fundamental character and
its beauty"** Fölsing, 108.

p. 103 **"...to make the mind tremble
with delight"** Judson, Horace Freeland,
The Search for Solutions. Quoted in
Ferris, 784–5.

p. 103 *I cannot find time to write...*
Calaprice, Alice (ed.), *The Expanded
Quotable Einstein.* Princeton: Princeton
University Press, 2000, 232.

p. 103 **"The breakthrough came
suddenly..."** Einstein, Albert, "How I
created the theory of relativity," trans.
Yoshimasa A. Ono. *Physics Today*
(August 1982): 47.

p. 104 **"the luckiest idea..."** Calaprice,
242.

p. 105 **"...the original relativity is
child's play"** quoted in Pais, Abraham,
*'Subtle is the Lord...': The Science and
the Life of Albert Einstein.* Oxford:
Oxford University Press, 1982, 216.

p. 105 **"...beautiful beyond comparison"** Calaprice, 234.

p. 106 **"fills me with great satisfaction"** quoted in Fölsing, 373.

p. 107 **"...but has never been repeated"**
Ibid., 381.

p. 107 **the work of a lone genius** Some
historians had claimed that German
mathematician David Hilbert
(1862–1943) had been the first to
develop the equations of general relativity. Recently, however, a team of
scholars investigating the original
documents concluded that Hilbert
took key ideas from one of Einstein's
manuscripts. See Correy, L., with J.
Renn and J. Stachel, "Belated decision
in the Hilbert-Einstein priority
dispute." *Science* 278 (5341) (1997):
1270–1273.

p. 108 **"one of the highest achievements..."** Fölsing, 444.

p. 109 **"...I would have to feel sorry for
God..."** quoted in Fölsing, 439.

p. 109 **"Revolution in Science"** *The Times* (London) (7 November 1919): 12.

p. 109 **"Lights All Askew in the Heavens"** *The New York Times* (10 November 1919): 17.

p. 109 **"...every coachman and every waiter..."** Calaprice, 238.

p. 110 **"saved the Swedish Academy..."** Fölsing, 536.

p. 110 **"cannot profess to follow..."** "Dr. Einstein's Theory." *The Times* (London) (28 November 1919): 13.

p. 110 **as well as a popular book** A current edition is Einstein, Albert, *Relativity: The Special and the General Theory.* New York: Crown Trade Paperbacks, 1961 (2nd ed.).

p. 111 **"lonely old song"** quoted in Fölsing, 553.

p. 111 **"Even devoted admirers of Einstein..."** Fölsing, 553.

p. 112 **confirmed by many different tests** For a detailed account of tests of general relativity through the mid-1980s, see Will, Clifford, *Was Einstein Right?* New York: Basic Books, 1986.

p. 113 **gravity is such a weak force** We think of gravity as being strong only because its effects can be felt over large distances. With electromagnetism, the effect of positive charges is almost always balanced by the effects of negative charges; with gravity, there is no such cancellation. Mass always attracts mass. Nonetheless, electromagnetism is the stronger force. In the case of two protons, for example, their electromagnetic repulsion is larger than their gravitational attraction by a factor of 10^{39}.

p. 114 **The universe was indeed expanding** This does not mean t the Milky Way is the centre of the universe. An observer in another gal. would see the same effect. A comparison is often made to a balloon with a number of dots drawn on its surface with a black marker. As the balloon expands, each dot moves away from every other dot.

p. 115 **a mysterious force is...pushing galaxies apart** See, for example, Krauss, Lawrence M., "Cosmological Antigravity." *Scientific American* (January 1999): 53–59.

p. 116 **"...the smallest possible number of hypotheses..."** Einstein, Albert, *Ideas and Opinions.* New York: Dell Publishing Co., 1954, 275.

Quantum Theory and Modern Physics

p. 117 *Anyone who is not shocked...* quoted in Gribbin, John, *In Search of Schrödinger's Cat: Quantum Physics and Reality.* New York: Bantam Books, 1984 (1988 ed.), 5.

p. 117 *To make a discovery...* Pais, Abraham, *Inward Bound: Of Matter and Forces in the Physical World.* New York: Oxford University Press, 1986, 134.

p. 119 **"All bodies of sensible magnitude..."** quoted in Purrington, Robert D., *Physics in the Nineteenth Century.* New Brunswick, NJ: Rutgers University Press, 1997, 119.

p. 120 **"...as if you fired a fifteen-inch shell..."** quoted in Pais, 189.

p. 121-122 **"...before breakfast, and still be hungry"** quoted in Pais, 190.

...ch as an electron orbiting a ...s In the case of the atom, the ...ron's speed may not be changing, ... its direction of motion, and there-ore its velocity, would be—and any change in velocity implies an acceleration.

p. 122 **"a family of ministers and lawyers..."** Crease, Robert P. and Charles C. Mann, *The Second Creation: Makers of the Revolutions in Twentieth-Century Physics*. New York: MacMillan Publishing Company, 1986, 23.

p. 123 **"an act of desperation"** quoted in Gribbin, 51.

p. 126 **one hundredth of one per cent** Spielberg, Nathan and Bryon D. Anderson, *Seven Ideas That Shook the Universe*. New York: John Wiley & Sons, Inc., 1987, 203.

p. 126 **The development of quantum mechanics...** Guillemin, Victor, *The Story of Quantum Mechanics*. New York: Charles Scribner's Sons, 1968, 102.

p. 127 **multiplying a particle's mass by its speed** I have simplified the picture slightly. In classical mechanics, momentum is mass times velocity (not speed); the difference is that velocity has a direction as well as a magnitude. For fast-moving objects, the formula has to be modified slightly, as Einstein showed with his special theory of relativity.

p. 128 **"...is again very muddled"** quoted in Gribbin, 99.

p. 128 **"...this wealth of mathematical structures..."** quoted in Gribbin, 103.

p. 131 **"...that He is not playing at dice"** quoted in Pais, 261.

p. 132 **"...throws the dice on every possible occasion"** Hawking, Stephen, Lecture at the Field Museum of Natural History, Chicago. 17 December 1996. Personal tape recording.

p. 132 **...of the scientist who knows the absolute truth** Reichenbach, Hans, The Rise of Scientific Philosophy. As quoted in Ayer, A.J. and Jane O'Grady, *A Dictionary of Philosophical Quotations*. Malden, MA: Blackwell Publishers Ltd., 1992 (1994 ed.), 371.

p. 133 **"...I reach for my gun."** Hawking, speaking to author Timothy Ferris, was paraphrasing the Nazi leader Hermann Goering, who is reported to have said, "Whenever I hear the word 'culture,' I reach for my revolver." See Ferris, Timothy, *The Whole Shebang: A State-of-the-Universe(s) Report*. New York: Simon & Schuster, 1998, 276, 345.

p. 135 **physicists at a laboratory in Colorado** Myatt, C.J. et al., "Decoherence of quantum superpositions through coupling to engineered reservoirs." *Nature* 403(6767) (2000): 269–273.

p. 136 **no matter how far apart they are** Pool, Robert, "Score one (more) for the spooks." *Discover* (January 1998): 53. This instantaneous correlation between distant particles may sound like it violates Einstein's special theory of relativity, which prohibits motion faster than the speed of light. In fact, quantum entanglement is consistent with special relativity, though the reasons are rather technical. See, for example, Gribbin, 228–229.

p. 136 **by the French physicist Alain Aspect** Aspect, Alain, Jean Dalibard, and Gerard Roger, "Experimental test of Bell's inequalities using time-varying analyzers." *Physical Review Letters* 49 (25) (1982): 1804–1807.

p. 136 **researchers at the University of Geneva** Pool, 53

p. 136 **"..whether we like it or not"** Tegmark, Max and John Wheeler, "100 years of quantum mysteries." *Scientific American* (February 2001): 72.

p. 138 **a modest 125 on an IQ test** Simmons, John, *The Giant Book of Scientists: The 100 Greatest Minds of All Time.* Sydney: The Book Company, 1996, 247.

p. 138 **"If that's the world's smartest man..."** quoted in Gleick, James, *Genius: The Life and Science of Richard Feynman.* New York: Pantheon Books, 1992, 397.

p. 138 **no more than a hundredth of a millimeter** Kleppner, Daniel and Roman Jackiw, "One hundred years of quantum physics." *Science* 289 (5481) (2000): 898.

p. 141 **The Standard Model embraces...18 particles** Again I have simplified things slightly. Each of the fermions is also thought to have a corresponding antiparticle possessing the same mass but the opposite electric charge. The number of entities embraced by the Standard Model depends on how one tallies these particles and their antiparticles.

p. 143-144 **"...the patchwork quilt has become a tapestry"** Glashow, Sheldon, The Nobel Foundation website, 27 February 2002 <www.nobel.se/physics/laureats/ 1979/glashow-lecture.html>

p. 144-145 **most successful the** **the history of science** Kleppner, «

p. 145 **one-quarter of the economi** **of the industrialized world** Lederma. Leon with Dick Teresi, *The God Particle: If the Universe is the Answer, What is the Question?* New York: Bantam Doubleday Dell Publishing, 1994, 185.

p. 145 **may eventually be used to build a "quantum computer"** See, for example, Seife, Charles, "The quandary of quantum information." *Science* 293 (5537) (2001): 2026–2027.

p. 146 **"...no longer within the grasp of the intelligent amateur..."** Barzun, Jacques, *From Dawn to Decadence, 1500 to the Present: 500 Years of Western Cultural Life.* New York: Harper Collins, 2000 (2001 ed.), 750.

p. 147 **"...order emerges from what appears to be chaos"** Planck, Max, *Where is Science Going?* London: Unwin Brothers Ltd., 1933, 13.

p. 148 **"...too byzantine, to be the full story"** Llewellyn Smith, Chris, "The large hadron collider." *Scientific American* (July 2000): 72.

p. 148 **"...clearly not the final answer"** Weinberg, Steven, "The great reduction: Physics in the twentieth century." In Howard, Michael and W. Roger Louis (eds.), *The Oxford History of the Twentieth Century.* New York: Oxford University Press, 1998, 33.

p. 148 **"...the theory's ultimate undoing"** Greenberger, Daniel and Anton Zeilinger. "Quantum theory: still crazy after all these years." *Physics World* (September 1995): 38.

p Loose Ends

From all things the one... quoted ...eale, Giovanni, *A History of Ancient ...ilosophy. Vol. 1. From the Origins to Socrates* (trans. John R. Catan). Albany, NY: State University of New York Press, 1987, 33.

p. 149 *Sooner or later we shall discover...* Weinberg, Steven, "The future of science and the universe." *New York Review of Books* (15 November 2001): 58.

p. 150 **"...we have the beginning of a new theory..."** Gross, David and Edward Witten, "The frontier of knowledge." *The Wall Street Journal* (12 July 1996): A12.

p. 154 **"...magic, mystery, or membrane..."** quoted in Duff, Michael J., "The theory formerly known as strings." *Scientific American* (February 1998): 64.

p. 154-155 **"widely regarded as the most gifted physicist..."** "*TIME*'s 25 Most Influential Americans." *TIME* (17 June 1996): 26.

p. 155 **"...Everything he does is golden"** The speaker was Harvard physicist Sidney Coleman. Quoted in Cole, K.C., "A theory of everything." *The New York Times Magazine* (18 October 1987): 20.

p. 155 **"...a vast subterranean archipelago of gold..."** Witten, Edward, Personal interview. 6 May 1997.

p. 155-156 **"...viable as a theory of quantum gravity..."** Peet, Amanda, Telephone interview. 21 August 2001.

p. 156 **"I don't know of any other contender..."** Davies, Paul, author interview for CBC Radio. 1 May 1997.

p. 156 **"over sold" and "pretty pathetic"** Hawking, Stephen and Roger Penrose, *The Nature of Space and Time.* Princeton: Princeton University Press, 1996, 4, 123.

p. 156 **"...like believing that God put fossils into the rocks..."** Hawking, Stephen, *The Universe in a Nutshell.* New York: Bantam Books, 2001, 57. (Witten had already been using the "fossil analogy" in his lectures.)

p. 156 *...even light can no longer escape* Hawking, Stephen, *The Universe in a Nutshell*, 111.

p. 156 *...that book by that wheelchair guy* From *The Simpsons Halloween Special VI*, 29 October 1995.

p. 158 **Hawking and Beckenstein worked out a formula** See, for example, Hawking, Stephen, *A Brief History of Time: From the Big Bang to Black Holes.* New York: Bantam Books, 1998 (1990 ed.), 99–113.

p. 158 **used string theory to count a black hole's quantum states** Strominger, Andrew and Cumrun Vafa, "Microscopic origin of the Beckenstein-Hawking entropy." *Physics Letters B* 379 (1996): 99–104.

p. 158-159 **"...something to say about the universe"** Peet, Amanda, Telephone interview. 21 August 2001.

p. 159 **"...one tends to believe it"** Hawking, Stephen, Lecture at the Field Museum of Natural History, Chicago. 17 December 1996. Personal tape recording.

p. 159 **"greatly increases our confidence..."** Hawking, Stephen, Lecture at the University of Toronto. 27 April 1998. Personal tape recording.

p. 159 *There is a theory which states...* Adams, Douglas, *The Restaurant at the End of the Universe*. London: Pan Books, 1980, 7–8.

p. 162 **The new theory was called** *inflation* Inflation solved a number of other problems as well. For a popular account, see Nadis, Steve, "Cosmic inflation comes of age." *Astronomy* (April 2002): 28–32.

p. 163 **known as the "ekpyrotic universe"** Steinhardt, Paul, Justin Khoury, Burt A. Ovurt, Nathan Sieberg, and Neil Turok, "Ekpyrotic universe: Colliding branes and the origin of the hot big bang." *Physical Review D* 64 (12) (2001): article 123572.

p. 163 **he proposes a "cyclic universe"** Steinhardt, Paul, Justin Khoury, Burt A. Ovurt, and Neil Turok, "From big crunch to big bang." Los Alamos electronic preprint archive, 22 March 2002 <xxx.lanl.gov/PS_cache/hep-th/pdf/01 08/0108187.pdf>. For a popular account, see Chown, Marcus, "Cycles of Creation." *New Scientist* (16 March 2002): 26–30.

p. 164 **"...I think string theory can answer this"** Steinhardt, Paul, Telephone interview. 7 February 2002.

p. 164 **"...who are talking to each other"** Peet, Amanda, Telephone interview. 21 August 2001.

p. 165 **"...or at least string-compatible"** Starkman, Glenn, Telephone interview. 22 August 2001.

p. 166 **"...It's the only game in town"** Ibid.

p. 166 **"to keep this contagious disease...out of Harvard"** Glashow, Sheldon, BBC Radio interview. Printed in Davies, P.C.W. and Julian Brown, *Superstrings: A Theory of Everything?* Cambridge: Cambridge University Press, 1988 (1995 ed.), 191.

p. 166 **"...It lives in the sky, so to speak"** Glashow, Sheldon, Personal interview. 8 May 1997.

p. 167 **The LHC should have enough clout** See, for example, Llewellyn Smith, Chris, "The large hadron collider." *Scientific American* (July 2000): 72–77.

p. 168 **"...to be the right theory of the world"** Peet, Amanda, Telephone interview. 21 August 2001.

p. 170 **"...question to be explored experimentally"** Schwarz, John, Telephone interview. 22 August 2001.

p. 171 **An intriguing variation** Randall, L. and R. Sundrum, "An alternative to compactification." *Physical Review Letters* 83 (23) (1999): 4690–4693.

p. 171 **"...within higher-dimensional space"** quoted in Chown, Marcus, "The great beyond." *New Scientist* (18 December 1999): 8.

p. 172 **"...opened up an entirely new branch..."** Randall, Lisa and Matthew D. Schwartz. "Unification and hierarchy from 5D anti-de sitter space." *Physical Review Letters* 88 (8) (2002): 081801-1.

p. 172 *Mathematics possesses not only truth...* quoted in Coxeter, H.S.M., ntroduction to Geometry. John Wiley & Sons, Inc., New York: 1961 (1966 ed.), xvii.

p. 172 "...a jungle red in tooth and claw" van Fraassen, B.C., The Scientific Image. New York: Clarendon Press, 1980, 40.

p. 173 "marvellous symmetry of the universe..." Copernicus, Nicolaus, De Revolutionibus Orbium Caelestium, trans. D.R. Danielson. In Danielson (ed.), The Book of the Cosmos: Imagining the Universe from Heraclitus to Hawking. Cambridge, MA: Helix, 2000, 117.

p. 173 "filled with unbelievable delight..." Baumgardt, Carola, Johannes Kepler: Life and Letters. New York: Philosophical Library, 1951, 121.

p. 173 "beautiful beyond comparison" Calaprice, Alice (ed.), The Expanded Quotable Einstein. Princeton: Princeton University Press, 2000, 234.

p. 173 "wealth of mathematical structures" quoted in Gribbin, John, In Search of Schrödinger's Cat: Quantum Physics and Reality. New York: Bantam Books, 1984 (1988 ed.), 103.

p. 173 "...more important to have beauty..." quoted in Judson, Horace Freeland, "The art of discovery." In Ferris, Timothy (ed.), The World Treasury of Physics, Astronomy, and Mathematics. New York: Little, Brown and Company, 1991, 786.

p. 173-174 "...most beautiful and consistent structure..." Vafa, Cumrun, Personal interview. 6 May 1997.

p. 174 "the only reliable source of truth" quoted in Fölsing, Albrecht,

Albert Einstein. New York: Penguin Books, 1998, 561.

p. 174 "...through mathematics alone" Fölsing, 561.

p. 175 "...we neither understand nor deserve" quoted in Ferris, Timothy (ed.), The World Treasury of Physics, Astronomy, and Mathematics. New York: Little, Brown and Company, 1991, 568, 540.

p. 175 "It seems to be a profound truth..." Schwarz, John, Telephone interview. 22 August 2001.

p. 175 "...the discovery of nature's final laws" Weinberg, Steven, Dreams of a Final Theory: The Scientist's Search for the Ultimate Laws of Nature. New York: Random House, 1992 (1994 ed.), 90.

p. 176 "...beautiful and simple in many ways" Lederman, Leon, Author interview for CBC Radio. 2 May 1997.

p. 177 "...odious column of bolted metal" Website of The Eiffel Tower "The Tower Stirs Debate and Controversy." 6 March 2002 <www.tour-eiffel.fr/teiffel/uk/docu-mentation/dossiers/page/debats.html>

p. 178 "...never conceived I would need to know" Schwarz, John, Telephone interview. 22 August 2001.

p. 178 "...the depths of our ignorance" Dyson, Freeman J., "Butterflies and superstrings." In Ferris, Timothy (ed.), The World Treasury of Physics, Astronomy, and Mathematics. New York: Little, Brown and Company, 1991, 129.

p. 178 "...I had lost most of it" Lederman, Leon with Dick Teresi, The

God Particle: If the Universe is the Answer, What is the Question? New York: Bantam Doubleday Dell Publishing, 1994, 394.

p. 179 "...had to travel to Europe" Witten, Edward, Personal interview. 6 May 1997.

p. 179 **more widespread and thus more familiar** This parallel between the aesthetics of the arts (including architecture) and the physical sciences was discussed recently in a thoughtful article by the philosopher James McAllister. He says that our perception of aesthetic beauty may be unconsciously dictated by a theory's empirical success: once we see that a theory is making correct predictions about nature, we come to view it as beautiful. See McAllister, James, "Is beauty a sign of truth in scientific theories?" *American Scientist* (March-April 1998); 174–183.

p. 179 "...this being a work in progress" Barrow, John, Author interview for CBC Radio. 13 May 1997.

p. 180 "...because it transcends its craft" Tolstoy, Ivan, *James Clerk Maxwell: A Biography.* Edinburgh: Canongate, 1981, 105.

p. 180 "...at the expense of all other areas" quoted in Glanz, James, "With little evidence, string theory gains influence." *The New York Times* (13 March 2001): D4.

p. 180 "...doesn't mean they are right" quoted in Dreifus, Claudia, "A mathematician at play in the fields of spacetime." *The New York Times* (19 January 1999): F3.

p. 181 "...modest about their 'breakthroughs'" 't Hooft, Gerard, E-mail interview. 6 February 2002.

p. 181 "...a gaping hole at the center of the M-theory..." Hawking, Stephen, *The Universe in a Nutshell,* 175.

p. 181 "...of this marvellous mathematical edifice" Schwarz, John, "Recent progress in superstring theory." Los Alamos electronic preprint archive, 8 March 2002 <xxx.lanl.gov/PS_cache/hep-th/pdf/00 07/0007130.pdf.>

p. 181 **into the hands of twentieth-century physicists** The phrase originated with the Italian physicist Daniele Amanti but was popularized by Witten.

p. 181 "...finding some advanced tool left behind..." quoted in Taubes, Gary, "A theory of everything takes shape." *Science* 269 (5230) (1995): 1513.

p. 182 "**The theorist should not be carped at...**" Einstein, Albert, *Ideas and Opinions.* New York: Dell Publishing Co., 1954, 275–6.

What Does It All Mean?

p. 183 *In reality we know nothing...* quoted in Taylor, C.C.W. (ed.), *The Routledge History of Philosophy.* Vol. I. New York: Routledge, 1997, 229.

p. 183 *The eternal mystery of the world...* Einstein, Albert, *Ideas and Opinions.* New York: Dell Publishing Co., 1954, 285.

p. 185 *Science cannot tell us...* Schrödinger, Erwin, *Mind and Matter.* Excerpted in Wilbur, Ken (ed.), *Quantum Questions: Mystical Writings of the World's Greatest Physicists.*

ton: Shambhala Publications, Inc., 84 (2001 ed.), 84.

p. 186 **"...to vote in the next election"** Davies, Paul, Author interview for CBC Radio. 1 May 1997.

p. 187 **"...a better video-game display"** Starkman, Glenn, Telephone interview. 22 August 2001.

p. 188 **"...no poet could make an epic out of them..."** Barzun, Jacques, *From Dawn to Decadence, 1500 to the Present: 500 Years of Western Cultural Life.* New York: Harper Collins, 2000 (2001 ed.), 750.

p. 189 **or John Haines' "Little Cosmic Dust Poem"** All three poems can be found in Ferris, Timothy (ed.), *The World Treasury of Physics, Astronomy, and Mathematics.* New York: Little, Brown and Company, 1991.

p. 191 **"...is scarcely worth kicking"** Eddington, A.S., *The Nature of the Physical World.* Cambridge: Cambridge University Press, 1928, 327.

p. 191 **"...what we can *say* about nature"** Bohr, Niels, *Essays on Atomic Physics and Human Knowledge 1932–1957.* Quoted in Pullman, Bernard, *The Atom in the History of Human Thought.* Oxford: Oxford University Press, 1998, 301.

p. 192 **"...confuse the image with reality"** Gribbin, John, *Schrödinger's Kittens and the Search for Reality.* New York: Little, Brown, and Company, 1995, 186.

p. 193 **"...as being any kind of 'ist.'"** Witten, Edward, Telephone interview. 8 Mar 2002.

p. 193 **"...as I am a positivist..."** Hawking, Stephen, *The Universe in a Nutshell.* New York: Bantam Books, 2001, 54.

p. 194 **"...but not to the facts themselves"** Brown, James Robert, *Who Rules in Science? An Opinionated Guide to the Wars.* Cambridge, MA: Harvard University Press, 2001, 3.

p. 194 **"The double standard...is glaring"** Brown, 176.

p. 194 **"These days it's fashionable..."** Davies, Paul, Telephone interview for CBC Radio. 16 August 2000.

p. 195 **"...believe the stars were really there"** quoted in Planck, Max, *Where is Science Going?* London: Unwin Brothers Ltd., 1933, 213.

p. 195 *Physics isn't a religion...* Lederman, Leon with Dick Teresi, *The God Particle: If the Universe is the Answer, What is the Question?* New York: Bantam Doubleday Dell Publishing, 1994.

p. 196 **"...providence of the Creator"** Copernicus, Nicolaus, *De Revolutionibus Orbium Caelestium,* trans. D.R. Danielson. In Danielson (ed.), *The Book of the Cosmos: Imagining the Universe from Heraclitus to Hawking.* Cambridge, MA: Perseus Publishing, 2000, 115.

p. 196 **"...the creation and its greatness"** Baumgardt, Carola, *Johannes Kepler: Life and Letters.* New York: Philosophical Library, 1951, 33.

p. 196 **"...may ascend to the starry firmament"** Somerville, Mary, *On the Connexion of the Physical Sciences.* In Danielson, 299.

p. 197-198 "**Science without religion is lame...**" Einstein, Albert, *Ideas and Opinions*, 55.

p. 198 "**My comprehension of God...**" Calaprice, Alice (ed.), *The Expanded Quotable Einstein.* Princeton: Princeton University Press, 2000, 203, 204.

p. 198 "**...extremely shaky or non-existent**" Lederman, 190.

p. 198 "**...for God to watch man's struggle...**" Feynman, Richard, *The Pleasure of Finding Things Out.* Cambridge, MA: Helix Books, 1999, 250.

p. 199 "**...the more it also seems pointless**" Weinberg, Steven, *The First Three Minutes: A Modern View of the Origin of the Universe.* New York: Harper Collins, 1977 (1988 ed.), 154.

p. 199 "**...I think that we will not**" Weinberg, Steven, *Dreams of a Final Theory*, 245.

p. 199 "**...Creator with a plan and an intention...**" Gingerich, Owen, "Let there be light: modern cosmogony and biblical creation." In Ferris, Timothy (ed.), *The World Treasury of Physics, Astronomy, and Mathematics.* New York: Little, Brown and Company, 1991, 386.

p. 199 **40 per cent of U.S. scientists** Larson, Edward J. and Larry Witham, "Scientists and religion in America." *Scientific American* (September 1999): 88–93.

p. 200 **members of the National Academy of Sciences** Angier, Natalie, "Confessions of a lonely atheist." *The New York Times Magazine* (14 January 2001): 37.

p. 200 "**timelessly guaranteeing of laws...**" Davies, telephone interv

p. 200 "**...very abstract notions of God...**" Andresen, Jensine, Personal interview. 16 December 1999.

p. 201 "**...We should not retreat from that accomplishment**" Weinberg was speaking at a conference in Washington, D.C. called "Cosmic Questions" in April 1999. The speech was later printed as "A designer universe?" in *The New York Review of Books* (21 October 1999), 48.

p. 201 "**know the mind of God**" Hawking, Stephen, *A Brief History of Time: From the Big Bang to Black Holes.* New York: Bantam Books, 1998 (1990 ed.), 175.

p. 201 "**like looking at God**" quoted in "Discovery bolsters big bang theory." *The Globe and Mail* (Toronto) (24 April 1992): A1.

p. 202 "**Christ is indispensable...**" Dembski, William, *Intelligent Design: The Bridge Between Science and Theology.* Downers Grove, IL: InterVarsity Press, 1999, 206, 210.

p. 202 **Kenneth Miller...in his recent book** Miller, Kenneth R., *Finding Darwin's God: A Scientist's Search for the Common Ground Between God and Evolution.* New York: Harper Collins, 1999.

p. 202 **fine-tuning problem** For a detailed discussion see Ferris, Timothy, *The Whole Shebang: A State-of-the-Universe(s) Report.* New York: Simon & Schuster, 1998, 297–306.

p. 203 "**...that it was all put together for us**" Ferris, Timothy, *The Whole Shebang*, 305.

"If I were granted omnipo-
...e..." Russell, Bertrand, *Religion and
...ence*. Oxford: Oxford University
ress, 1961 (1997 ed.), 222.

p. 203 **multiple universes** See, for
example, Rees, Martin, *Just Six
Numbers: The Deep Forces That Shape
the Universe*. New York: Basic Books,
1999 (2000 ed.).

p. 204 **"...not merely metaphysics"**
Rees, Martin, Author interview for
CBC Radio. 22 September 2000.

p. 204 **"...at the very deepest level"**
Rees, Martin, *Before the Beginning: Our
Universe and Others*. Reading, MA:
Helix Books, 1997 (1998 ed.), 253.

p. 204 *One God, one law...* Tennyson,
Alfred (Lord), "In Memoriam." In Hill,
Robert W., Jr. (ed.), *Tennyson's Poetry*.
New York: W.W. Norton & Company,
Inc., 1971.

p. 205 **"...something underpinning it
all..."** Davies, Paul, Telephone interview
for CBC Radio. 16 August 2000.

p. 205 **"...akin to a religious feeling..."**
Einstein, Albert, *Ideas and Opinions*.
New York: Dell Publishing Co., 1954,
255.

p. 205 **"...one God, as it were, rather
than many..."** Barrow, John, Author
interview for CBC Radio. 13 May 1997.

p. 206 **"...plays a role in one's thinking"** Salam spoke in a BBC Radio
interview. Quoted in Wolpert, Lewis
and Allison Richards, *A Passion for
Science*. Oxford: Oxford University
Press, 1988 (1989 ed.), 18.

p. 206 **"...part of the adaptive evolutionary process..."** Barrow, John,
Author interview.

p. 207 **"...no *reason* that things get
simpler..."** quoted in Crease, Robert P.
and Charles C. Mann, *The Second
Creation: Makers of the Revolutions in
Twentieth-Century Physics*. New York:
MacMillan Publishing Company, 1986,
417.

p. 207 **"...I would not be interested in
them"** quoted in Wheeler, John,
"Mercer Street and other memories." In
Ferris, Timothy (ed.), *The World
Treasury of Physics, Astronomy, and
Mathematics*. New York: Little, Brown
and Company, 1991, 566.

Epilogue

p. **209 *The answer...is Forty-two***
Adams, Douglas, *The Hitch Hiker's
Guide to the Galaxy*. London: Pan
Books, 1979 (1983 ed.), 135.

p. 211 **"...graduate students today..."**
Witten, Edward, Telephone interview. 8
March 2002.

p. 211 **"...and ever more rigorous
tests"** Popper, Karl, "The logic of scientific discovery." Quoted in Ferris,
Timothy (ed.), *The World Treasury of
Physics, Astronomy, and Mathematics*.
New York: Little, Brown and Company,
1991, 800.

p. 211 **"...subject to question and
doubt"** Einstein, Albert, *Ideas and
Opinions*, 315.

p. 212 **"...part of your belief system"**
Lederman, Leon, Author interview for
CBC Radio. 2 May 1997.

p. 212 **"But I think it will happen..."**
Weinberg, Steven, "The future of
science, and the universe." *The New
York Review of Books* (15 November
2001): 58.

p. 212 "...by 2010, 2015 at the outside."
Smolin, Lee, *Three Roads to Quantum Gravity*. London: Weidenfeld & Nicholson, 2000 (Basic Books edition, 2001), 211.

p. 213 "Encouraged...by the fact that *science works*" Lederman interview, 2 May 1997.

p. 213 "...I hope that we'll live to see it" Witten, Edward, Personal interview. 6 May 1997.

p. 214 "...would be very nice to discover" Feynman, Richard, *The Pleasure of Finding Things Out*. Cambridge, MA: Helix Books, 1999, 23.

p. 215 "Narrow are the powers..."
Kirk, G.S., J.E. Raven, and M. Schofield, *The Presocratic Philosophers: A Critical History with a Selection of Texts*. Cambridge: Cambridge University Press, 1957 (1987 ed.), 285.

Credits

Quotations from *The Life of Isaac Newton* by Richard Westfall (1994), *The Lord of Uraniborg* by Victor E. Thoren (1990), *On Tycho's Island: Tycho Brahe and his Assistants, 1570-1601* by John Robert Christianson (2000), and *The Presocratic Philosophers: A Critical History With a Selection of Texts* (1957; 2nd ed. 1983) by G.S. Kirk, J.E. Raven, and M. Schofield are reprinted with the permission of Cambridge University Press.

Quotations from *Isaac Newton—The Principia: A New Translation* translated and edited by I. Bernard Cohen and Anne Whitman (1999) are used courtesy of the University of California Press and the Regents of the University of California.

Quotations from *Dreams of a Final Theory* by Steven Weinberg (1992), *Discoveries and Opinions of Galileo* by Stillman Drake (Anchor edition 1957), and *The Universe in a Nutshell* by Stephen Hawking (2001) are used courtesy of Random House.

Quotations from *De Revolutionibus Orbium Caelestium* by Nicolaus Copernicus, translated by Dennis Richard Danielson in *The Book of the Cosmos: Imagining the Universe from Heraclitus to Hawking* (Cambridge Mass.: Perseus Publishing 2000) are used by permission of the translator.

Quotations from *Revolution in Science* by I. Bernard Cohen (Cambridge Mass.: The Belknap Press of Harvard University Press 1985) are reprinted by permission of the publisher and the President and Fellows of Harvard College.

Quotations from *Albert Einstein* by Albrecht Fölsing (Penguin Putnam 1998) and *The Origins of Scientific Thought* by Georgio de Santillana (New American Library edition 1961) are used courtesy of Penguin Putnam Inc.

Quotations from *From Dawn to Decadence, 1500 to the Present: 500 Years of Western Cultural Life* by Jacques Barzun (2000) are used courtesy of HarperCollins.

Quotations from *James Clerk Maxwell: A Biography* by Ivan Tolstoy (Canongate 1991) are used courtesy of the author.

Quotation from *Schrödinger's Kittens and the Search for Reality* by John Gribbin (1995) is used courtesy of Little, Brown and Company.

Quotation from *Albert Einstein: A Biographical Memoir* by John A. Wheeler (1980) is used courtesy of National Academy Press.

Quotation from "Quantum theory: still crazy after all these years," by Daniel Greenberger and Anton Zeilinger, in *Physics World* September 1995, page 38, is used courtesy of *Physics World*.

Quotation from "How I Created the Theory of Relativity," by Albert Einstein, translated by Yoshimasa A. Ono, *Physics Today* August 1982, page 47, as well as the Schrödinger's Cat drawing, are used courtesy of *Physics Today*.

Quotations from *Ideas and Opinions*, by Albert Einstein © 1954, 1982 Crown Publishers, Inc. are used by permission of Crown Publishers, a division of Random House, Inc.

Quotations from *The Expanded Quotable Einstein*, ed. Alice Calaprice (2000) are reprinted by permission of Princeton University Press.

Thanks to the Barbara Wolff and the Albert Einstein Archives of the Hebrew University of Jerusalem for assistance in verifying Einstein quotations.

Quotations from John Barrow, Paul Davies, Owen Gingerich, Leon Lederman, Amanda Peet, John Schwarz, Glenn Starkman, Steven Weinberg, and Edward Witten are used by permission.

Quotations from *Early Greek Philosophy* by Jonathan Barnes (1987) and *Herodotus—The Histories*, translated by Aubrey de Selincourt (1996) are used by permission of Penguin U.K.

Images from the American Institute of Physics are used courtesy of the Emilio Segrè Visual Archives; thanks to the Friends of the Center for History of Physics for their support of the collections. Louis de Broglie by A. Bortzells Tryckeri, Weber Collection; Albert Einstein with Niels Bohr by Paul Ehrenfest, Ehrenfest Collection; Stephen Hawking and Erwin Schrödinger from the Physics Today Collection; James Clerk Maxwell from the collection of Sir Henry Roscoe; Max Planck from the W.F. Meggers Gallery of Nobel Laureates.

Cartoons appear courtesy of The New Yorker Collection and cartoonbank.com. "five basic elements..." ©2000 J.P. Rini; "This is Merlin..." ©1994 Dana Fradon; "It's all string theory..." ©1998 Victoria Roberts; "If only it were so simple" ©1987 Bernard Schoenbaum; "Scientists confirm..."©1998 Jack Ziegler.

Photo of David Gross by Len Wood reprinted with permission from the Santa Barbara News-Press.

Thanks to Andrew Skelly, Tom Venetis, and Ann Venetis for author photo and "waves" photo.

Original artwork by Dave McKay.

Index

reene, Brian, 157, 174
Greenberger, Daniel, 148
Gribbin, John, 192
Gross, David, 150, 156
Grosseteste, Robert, 32
Guillemin, Victor, 126
Guth, Alan, 162
Haines, John, 189
Halley, Edmond, 66, 67, 72
Halley's Comet, 67, 72
Harvey, Jeff, 181
Hawking, Stephen, 67, 132, 133, 137, 156, 157-158, 181, 193, 201
Heisenberg, Werner, 126, 128, 130, 137, 173, 189
heliocentric model. *See* solar system
Henry, Joseph, 80
Heraclitus of Ephesus, 14, 149
Herodotus, 8-9, 11, 13
Hertz, Heinrich, 85, 102
Higgs, Peter, 142
Higgs boson, 142
Hobbes, Thomas, 10, 60
Hobsbawm, Eric, 89
Hubble, Edwin, 114
Hulse, Russell, 113
Hume, David, 134, 193
Huygens, Christiaan, 84
inverse-square law, 66, 74, 76, 169-170
John the Grammarian, 52
John Paul II (Pope), 61
Johnson, Samuel, 191, 193
Joshua, Book of, 31, 57-58
Judson, Horace, 103
inflation (in cosmology), 162-163
"Intelligent Design", 201-202
Kaluza-Klein theory, 168-169
Kant, Immanuel, 45, 71, 193, 206
Kepler, Johannes, 35, 36, 40, 41-47, 53, 48, 53, 55, 57, 63, 66, 77, 82, 173, 174, 193, 196, 201

Kuhn, Thomas, 172
Lederman, Leon, 1, 52, 145, 176, 177, 178, 195, 198, 212-213, 214
Leibniz, Gottfried, 67
Leucippus, 17, 118
light, nature of, 83-85, 90, 95, 124-125; speed of, 15, 85, 91-92, 96, 99-100, 210
Linde, Andrei, 164, 203
Lippershey, Hans, 53
Llewellyn Smith, Chris, 148
Long, A.A., 23
loop quantum gravity, 165
M-theory, 154, 156, 179, 181, 182, 187, 195
magnetism, 46, 72-73, 77-78, 87, 90, 208
Magnus, Albertus, 32
many-worlds. *See* quantum theory
Marconi, Guglielmo, 85
Mastlin, Michael, 41, 42, 44
materialism (philosophy of nature), 9, 19
mathematics, role of, in science, 21, 61-62, 70, 175-180
Maxwell, James Clerk, 6, 78, 81-85, 86, 90, 91, 92, 95, 96, 102, 113, 122, 123, 124, 125, 139, 179, 180
Maxwell's equations, 82, 86, 179
medieval science, 29-32, 72
Mendeleyev, Dmitri, 119
Menuhin, Yehudi, 89
Michelson-Morley experiment, 92
Mill, John Stuart, 193
Miller, Kenneth, 202
Milton, John, 60
Minkowski, Hermann, 88
models (in physics), 190-195
Morrison, David, 157
Morse, Samuel, 80
Muslim and Arab influence on science, 27-29, 199, 206

Napoleon, 74

Newton, Isaac, 6, 46, 49, 62-70, 72, 77, 82, 83, 84, 87, 88, 89, 90, 92, 97, 104, 108, 118, 125, 139, 146, 169, 170, 176, 177, 178, 196, 200; Cambridge, years at, 63-64, 66-67; discovery of universal gravitation, 64-66; early life, 62-63; interest in alchemy (vs. chemistry), 68-70; London, living in, 67-68; publication of *Principia*, 66-67

neutron, 121, 138, 139, 152, 187

nucleus, 121, 122, 139, 140, 150, 151, 152

Ockham, William of, 3, 5, 32

Ockham's Razor, 3, 32, 135

Oersted, Hans Christian, 75-77, 81, 86, 173

Pais, Abraham, 117

parallax, 35, 39

Pauli, Wolfgang, 128

Pecham, John, 32

Peet, Amanda, 156, 158-159, 164, 168

Penrose, Roger, 180

photoelectric effect, 95, 110, 124

photon, 124-125, 126, 134, 136, 137, 139, 141, 188

Planck, Max, 95, 122-124, 146, 147

Planck's constant, 123, 125, 129

planets. *See* solar system

Plato, 21, 26, 190-191, 192, 196

Polkinghorne, John, 199

Popper, Karl, 211

Presocratics, 12, 20, 187, 196

Priestley, Joseph, 74

Primack, Joel, 199

proton, 121, 138-139, 150, 152, 187

Ptolemy, 22-24, 30-31, 33-34, 36, 38, 39, 40, 55, 57, 174

Pullman, Bernard, 24

Pythagoras, 11

quantum chromoynamics, 140

quantum computer, 145

quantum electrodynamics, 138

quantum entanglement, 135-136, 145

quantum field theories, 137, 140, 147

quantum mechanics. *See* quantum theory

quantum theory, 111, 116, 118-148, 150, 151, 158, 159, 161, 179, 182, 187, 211; atomic structure, and, 125-126; early development of, 122-124; interpretations of, 134-135; technological applications of, 145-146

quark, 140, 141, 151, 184, 187, 188, 189, 193, 215

radio waves, 85

Randall, Lisa, 171-172

reductionism, 185-186

Rees, Martin, 203-204

Reichenbach, Hans, 132

relativity. *See* Einstein; general relativity; special relativity

religion, 78, 83, 195-202; Greek science and, 12-13, 19; influence on search for Theory of Everything, 204-206. *See also* Christianity; God

Reston, James, 55, 60, 61

Riemann, G.F. Bernhard, 104, 177

Roman Empire, 26-27

Royal Institution, 77

Royal Society, 67, 108

Rubbia, Carlo, 60

Rudolph II (Emperor), 40

Russell, Bertrand, 203

Rutherford, Ernest, 120-122, 131, 138, 191

Ryan, Richard, 189

Salam, Abdus, 143, 199, 206

Scherk, Joel, 151

Schrödinger, Erwin, 23, 126, 128, 130, 132, 137, 185

Schrödinger's Cat, 132-133, 135

Schrödinger equation, 129, 147

Schwarz, John, 151, 170, 174, 175, 178, 181

Schwinger, Julian, 137